Corrosion Protection against Carbon Dioxide

Edited by
Michael Schütze, Bernd Isecke, and Roman Bender

Corrosion Protection against Carbon Dioxide

WILEY-VCH Verlag GmbH & Co. KGaA

Editors:

Prof. Dr.-Ing. Michael Schütze
Director Materials
Karl Winnacker Institute of DECHEMA e. V.
Society for Chemical Engineering and Biotechnology
Theodor-Heuss-Allee 25
60486 Frankfurt (Main)
Germany

Prof. Dr.-Ing. Bernd Isecke
Head of the Department
Materials Protection and Surface Technologies
Federal Institute for Materials Research and Testing
Unter den Eichen 44–46
12203 Berlin
Germany

Dr. rer. nat. Roman Bender
Chief Executive of GfKORR e.V.
Society for Corrosion Protection
Theodor-Heuss-Allee 25
60486 Frankfurt (Main)
Germany

Cover Illustration
Source: Karl Winnacker Institute of DECHEMA e. V., Frankfurt (Main), Germany

Warranty Disclaimer

This book has been compiled from literature data with the greatest possible care and attention. The statements made only provide general descriptions and information.

Even for the correct selection of materials and correct processing, corrosive attack cannot be excluded in a corrosion system as it may be caused by previously unknown critical conditions and influencing factors or subsequently modified operating conditions.

No guarantee can be given for the chemical stability of the plant or equipment. Therefore, the given information and recommendations do not include any statements, from which warranty claims can be derived with respect to DECHEMA e. V. or its employees or the authors.

The DECHEMA e. V. is liable to the customer, irrespective of the legal grounds, for intentional or grossly negligent damage caused by their legal representatives or vicarious agents.

For a case of slight negligence, liability is limited to the infringement of essential contractual obligations (cardinal obligations). DECHEMA e. V. is not liable in the case of slight negligence for collateral damage or consequential damage as well as for damage that results from interruptions in the operations or delays which may arise from the deployment of the Handbook.

This book was carefully produced. Nevertheless, editors, authors and publisher do not warrant the information contained therein to be free of errors. Readers are advised to keep in mind that statements, data, illustrations, procedural details or other items may inadvertently be inaccurate.

Library of Congress Card No.: Applied for.

British Library Cataloguing-in-Publication Data:
A catalogue record for this book is available from the British Library.

Bibliographic information published by Die Deutsche Bibliothek
Die Deutsche Bibliothek lists this publication in the Deutsche Nationalbibliografie; detailed bibliographic data is available in the Internet at <http://dnb.ddb.de>.

© 2011 DECHEMA e. V., Society for Chemical Engineering and Biotechnology, 60486 Frankfurt (Main), Germany

All rights reserved (including those of translation into other languages). No part of this book may be reproduced in any form – nor transmitted or translated into machine language without written permission from the publishers. Registered names, trademarks, etc. used in this book, even when not specifically marked as such, are not to be considered unprotected by law.

Typesetting Kühn & Weyh, Freiburg
Printing Strauss GmbH, Mörlenbach
Binding Strauss GmbH, Mörlenbach
Cover Design Grafik-Design Schulz, Fußgönheim

Printed in the Federal Republic of Germany
Printed on acid-free paper

Print ISBN: 978-3-527-33145-1

Contents

Preface *VII*

How to use the Handbook *IX*

Carbon dioxide *1*
Authors: P. Drodten, D. Schedlitzki/Editor: R. Bender

A Metallic materials   *30*
   Aluminium and aluminium alloys, copper and copper alloys, iron, iron-based alloys and steels, nickel and nickel alloys, titanium and titanium alloys, zinc, cadmium and their alloys

B Nonmetallic inorganic materials   *100*
   Carbon and graphite, binders for building materials, glass, quartz ware and quartz glass, enamel, oxide ceramic materials, metal ceramic materials

C Organic materials/plastics   *112*
   Thermoplastics, thermosetting plastics, elastomers, duroplasts

D Materials with special properties   *148*
   Coatings and linings, seals and packings, composite materials

E Materials recommendations   *182*

Bibliography   *192*

Index of materials   *209*

Subject index   *216*

Preface

Practically all industries face the problem of corrosion - from the micro-scale of components for the electronics industries to the macro-scale of those for the chemical and construction industries. This explains why the overall costs of corrosion still amount to about 2 to 4 % of the gross national product of industrialized countries despite the fact that zillions of dollars have been spent on corrosion research during the last few decades.

Much of this research was necessary due to the development of new technologies, materials and products, but it is no secret that a considerable number of failures in technology nowadays could, to a significant extent, be avoided if existing knowledge were used properly. This fact is particularly true in the field of corrosion and corrosion protection. Here, a wealth of information exists, but unfortunately in most cases it is scattered over many different information sources. However, as far back as 1953, an initiative was launched in Germany to compile an information system from the existing knowledge of corrosion and to complement this information with commentaries and interpretations by corrosion experts. The information system, entitled "DECHEMA-WERKSTOFF-TABELLE" (DECHEMA Corrosion Data Sheets), grew rapidly in size and content during the following years and soon became an indispensable tool for all engineers and scientists dealing with corrosion problems. This tool is still a living system today: it is continuously revised and updated by corrosion experts and thus represents a unique source of information. Currently, it comprises more than 8,000 pages with approximately 110,000 corrosion systems (i.e., all relevant commercial materials and media), based on the evaluation of over 100,000 scientific and technical articles which are referenced in the database.

Last century, an increasing demand for an English version of the DECHEMA-WERKSTOFF-TABELLE arose in the 80s; accordingly the first volume of the DECHEMA Corrosion Handbook was published in 1987. This was a slightly condensed version of the German edition and comprised 12 volumes. Before long, this handbook had spread all over the world and become a standard tool in countless laboratories outside Germany. The second edition of the DECHEMA Corrosion Handbook was published in 2004. Together the two editions covered 24 volumes.

The present handbook compiles all information on the corrosion behaviour of materials that are in contact with carbon dioxide or environments containing this compound. This compilation is an indispensable tool for all engineers and scientists dealing with corrosion problems in carbon dioxide containing environments of any industrial use.

Carbon dioxide is used by many industrial sectors including the food industry, the oil industry, and the chemical industry. It is used in many consumer products that require pressurized gas because it is inexpensive and nonflammable, and because it undergoes a phase transition from gas to liquid at room temperature at an attainable pressure of approximately 60 bar, allowing far more carbon dioxide to fit in a given container than otherwise would. There is a wide range of areas of application. Most prominent usage includes pneumatic systems, fire extinguishers, welding, Pharmaceutical and other chemical processing, agricultural and biological applications, coal bed methane or oil bed recovery, refrigrants, lasers or pH-control. And last but not least Carbon dioxide is the most important greenhouse gas produced by human activities, primarily through the combustion of fossil fuels.

Corrosion is a complex phenomenon that depends on a number of parameters, related to both the environment and the metal. In this handbook the behaviour of materials in contact with carbon dioxide containing gases and liquids is compiled.

The chapters are arranged by the agents leading to individual corrosion reactions, and a vast number of materials are presented in terms of their behaviour in these agents. The key information consists of quantitative data on corrosion rates coupled with commentaries on the background and mechanisms of corrosion behind these data, together with the dependencies on secondary parameters, such as flow-rate, pH, temperature, etc. This information is complemented by more detailed annotations where necessary, and by an immense number of references listed at the end of the handbook.

An important feature of this handbook is that the data was compiled for industrial use. Therefore, particularly for those working in industrial laboratories or for industrial clients, the book will be an invaluable source of rapid information for day-to-day problem solving. The handbook will have fulfilled its task if it helps to avoid the failures and problems caused by corrosion simply by providing a comprehensive source of information summarizing the present state-of-the-art. Last but not least, in cases where this knowledge is applied, there is a good chance of decreasing the costs of corrosion significantly.

Finally the editors would like to express their appreciation to Lieselotte Wolf and Tina Kemmerich for their admirable commitment and meticulous editing of a work that is encyclopedic in scope.

They are also indebted to Gudrun Walter of Wiley-VCH for their valuable assistance during all stages of the preparation of this book.

Michael Schütze, Bernd Isecke and Roman Bender

How to use the Handbook

The Handbook provides information on the chemical resistance and the corrosion behavior of materials in carbon dioxide respectively carbonic acid.

The user is given information on the range of applications and corrosion protection measures for metallic, non-metallic inorganic, and organic materials, including plastics.

Research results and operating experience reported by experts allow recommendations to be made for the selection of materials and to provide assistance in the assessment of damage.

The objective is to offer a comprehensive and concise description of the behavior of the different materials in contact with the medium.

The book is subdivided according to four groups of materials A-D:

- A Metallic materials
- B Non-metallic inorganic materials
- C Organic materials and plastics
- D Materials with special properties

These material groups are each subdivided according to their chemical formula, the metals are classed according to different alloy groups. These groups are shown in the uniformly designed overview table at the start of each chapter.

The information on resistance is given as text, tables, and figures. The literature used by the authors is cited at the corresponding point. There is an index of materials as well as a subject index at the end of the book so that the user can quickly find the information given for a particular keyword.

The Handbook is thus a guide that leads the reader to materials that have already been used in certain cases, that can be used or that are not suitable owing to their lack of resistance.

The resistance is coded with three evaluation symbols in order to compress the information. Uniform corrosion is evaluated according to the following criteria:

Symbol	Meaning	Area-related mass loss rate[1]		Corrosion rate
		g/(m²h)	g/(m²d)	mm/a
+	resistant	≤ 0.1	≤ 2.4	≤ 0.1[2]
⊕	fairly resistant	< 1.0	< 24.0	< 1.0
−	not resistant	> 1.0	> 24.0	> 1.0

[1] For Al, Mg, and its alloys, 1/3 of the value must be used
[2] The values for Ta, Ti, and Zr are too high (possible embrittlement due to hydrogen absorption in the event of corrosion! Therefore, corrosion rate = 0.01 mm/a, see the individual cases)

The evaluation of the corrosion resistance of metallic materials is given

- for uniform corrosion or local penetration rate, in: mm/a
- or if the density of the material is not known, in: g/(m²h) or g/(m²d).

Pitting corrosion, crevice corrosion, and stress corrosion cracking or non-uniform attack are particularly highlighted.

The following equations are used to convert mass loss rates, x, into the corrosion rate, y:

from x_1 into g/(m²h)   from x_2 into g/(m²d)   where

$$\frac{x_1 \cdot 365 \cdot 24}{\rho \cdot 1000} = y \, (mm/a) \qquad \frac{x_2 \cdot 365}{\rho \cdot 1000} = y \, (mm/a)$$

x_1: value in g/(m²h)
x_2: value in g/(m²d)
ρ: density of material in g/cm³
y: value in (mm/a)
d: days
h: hours

In those media in which uniform corrosion can be expected, if possible, isocorrosion curves (corrosion rate = 0.1 mm/a) or resistance ranges for non-metallic materials are given. The evaluation criteria for non-metallic inorganic materials are stated in the individual cases; depending on the material and medium, they may also be given as corrosion rates (mm/a).

The suitability of organic materials is generally evaluated by comparing property characteristics (e.g. mass, tensile strength, elasticity module or ultimate elongation) and other changes (e.g. cracking) after exposure to the medium with respect to these characteristics in the initial state before exposure. The extent of changes in the properties after exposure to the medium is decisive for the evaluation of the resistance to chemicals or the durability of the materials. The criteria listed below for the evaluation of the chemical resistance apply to thermoplastics used to manufacture pipes and are based on results from immersion tests with an immersion time of 112 days (see ISO 4433 Part 1 to 4). In principle, they are also applicable to other organic materials; however, they should be adapted to the individual material, because, as the following table shows, the evaluation criteria are not consistent, even within a group of thermoplasts, but depend on the type of thermoplastic material.

Symbol	Meaning	Permissible limiting value[1]			
		of the mass change[2] %	of the tensile strength[3] %	of the elasticity module[3] %	of the ultimate elongation[3] %
+	resistant/ durable	PE, PP, PB: −2 to 10 PVC, PVDF: −0.8 to 3.6	PE, PP, PB, PVC, PVDF: ≥ 80	PE, PP, PB: ≥ 38 PVC: ≥ 83 PVDF: ≥ 43	PE, PP, PB: ≥ 50 to 200 PVC, PVDF: 50 to 125
⊕	limited resistance/ limited durability	PE, PP, PB: > 10 to 15 or < −2 to −5 PVC, PVDF: < −0.8 to −2 or > 3.6 to 10	PE, PP, PB, PVC, PVDF: < 80 to 46	PE, PP, PB: < 38 to 31 PVC: < 83 to 46 PVDF: < 43 to 30	PE, PP, PB: < 50 to 30 or > 200 to 300 PVC, PVDF: < 50 to 30 or > 125 to 150
−	not resistant/ not durable	PE, PP, PB: < −5 or > 15 PVC, PVDF: < −2 or > 10	PE, PP, PB, PVC, PVDF: < 46	PE, PP, PB: < 31 PVC: < 46 PVDF: < 30	PE, PP, PB: < 30 or > 300 PVC, PVDF: < 30 or > 150

[1] The data applies to the values determined in the initial state without exposure to the medium which corresponds to 100 %
[2] Relative mass change according to DIN EN ISO 175
[3] Tensile strength, elasticity module, and ultimate elongation according to DIN EN ISO 527-1
Scope of validity for PVC: PVC-U, PVC-HI, and PVC-C; for PE: PE-HD, PE-MD, PE-LD, and PE-X

Unless stated otherwise, the data was measured at atmospheric pressure and room temperature.

The resistance data should not be accepted by the user without question, and the materials for a particular purpose should not be regarded as the only ones that are suitable. To avoid wrong conclusions being drawn, it must be always taken into account that the expected material behavior depends on a variety of factors that are often difficult to recognize individually and which may not have been taken deliberately into account in the investigations upon which the data is based. Under certain circumstances, even slight deviations in the chemical composition of the medium, in the pressure, in the temperature or, for example, in the flow rate are sufficient to have a significant effect on the behavior of the materials. Furthermore, impurities in the medium or mixed media can result in a considerable increase in corrosion.

The composition or the pretreatment of the material itself can also be of decisive importance for its behavior. In this respect, welding should be mentioned. The suitability of the component's design with respect to corrosion is a further point which must be taken into account. In case of doubt, the corrosion resistance should be investigated under operating conditions to decide on the suitability of the selected materials.

Carbon dioxide

Author: P. Drodten, D. Schedlitzki/Editor: R. Bender

		Page
Survey Table		2
Preliminary remarks		4
A	**Metallic materials**	30
A 1	Silver and silver alloys	30
A 2	Aluminium	30
A 3	Aluminium alloys	32
A 4	Gold and gold alloys	32
A 5	Cobalt alloys	32
A 6	Chromium and chromium alloys	33
A 7	Copper	33
A 8	Copper-aluminium alloys	40
A 9	Copper-nickel alloys	40
A 10	Copper-tin alloys (bronze)	40
A 14	Unalloyed and low-alloy steels/cast steel	41
A 15	Unalloyed cast iron and low-alloy cast iron	41
A 17	Ferritic chrome steels with < 13 % Cr	67
A 18	Ferritic chrome steels with ≥ 13 % Cr	68
A 19	High-alloy multiphase steels	68
A 19.1	Ferritic/perlitic-martensitic steels/	68
A 20	Austenitic chromium-nickel steels	72
A 21	Austenitic CrNiMo(N) steels	89
A 22	Austenitic CrNiMoCu(N) steels	89
A 24	Magnesium and magnesium alloys	92
A 25	Molybdenum and molybdenum alloys	94
A 26	Nickel	94
A 27	Nickel-chromium alloys	95
A 28	Nickel-chromium-iron alloys (without Mo)	95
A 29	Nickel-chromium-molybdenum alloys	95
A 33	Lead and lead alloys	97
A 34	Platinum and platinum alloys	97
A 35	Platinum metals (Ir, Os, Pd, Rh, Ru) and their alloys	97
A 36	Tin and tin alloys	97
A 37	Tantalum, niobium and their alloys	97
A 38	Titanium and titanium alloys	97

		Page
A 39	Zinc, cadmium and their alloys	97
A 40	Zirconium and zirconium alloys	99
B	**Nonmetallic inorganic materials**	99
B 3	Carbon and graphite	100
B 4	Binders for building materials (e.g. concrete, mortar)	101
B 5	Acid-resistant building materials and binders (putties)	104
B 8	Enamel	105
B 12	Oxide ceramic materials	106
B 13	Metal-ceramic materials (carbides, nitrides)	106
C	**Organic materials/plastics**	112
	Thermoplastics	112
	Polyolefines and polyvinyl chlorides	112
	Fluoropolymers and high-temperature thermoplastics	119
	Polystyrene and copolymere	
	Polyester	
	Polyamides	
	Further thermoplastics	123
	Thermoplastic elastomers	135
	Duroplastics	137
	Epoxy resins	
	Unsaturated polyester resins	
	Vinyl ester resins	
	Polyurethanes	
	Phenolic resins	
	Furanic resins	
	Further duroplastics	
	Elastomers	139
D	**Materials with special properties**	148
D 1	Coatings and linings	148
D 2	Seals and packings	167
D 3	Composite materials	175
E	**Material recommendations**	182
Bibliography		192
Index of materials		209
Subject index		216

Survey Table

The corrosion behavior of the individual materials was evaluated on the basis of experience gained in practice and the conditions described in the following text.

Material No.*	Type	Behavior**
A	**Metallic materials**	
A 1	Silver and silver alloys	+
A 2	Aluminium	+ to −
A 3	Aluminium alloys	+ to −
A 4	Gold and gold alloys	+
A 5	Cobalt alloys	+
A 6	Chromium and chromium alloys	+
A 7	Copper	+ to −
A 8	Copper-aluminium alloys	+ to −
A 9	Copper-nickel alloys	+ to −
A 10	Copper-tin alloys (bronze)	+ to −
A 14	Unalloyed and low-alloy steels/cast steel	+ to −
A 15	Unalloyed cast iron and low-alloy cast iron	+ to −
A 17	Ferritic chrome steels with < 13 % Cr	+
A 18	Ferritic chrome steels with ≥ 13 % Cr	+ to ⊕
A 19	High-alloy multiphase steels	+ to ⊕
A 19.1	Ferritic/perlitic-martensitic steels/	+ to ⊕
A 20	Austenitic chromium-nickel steels	+
A 21	Austenitic CrNiMo(N) steels	+
A 22	Austenitic CrNiMoCu(N) steels	+
A 24	Magnesium and magnesium alloys	+ to −
A 25	Molybdenum and molybdenum alloys	+
A 26	Nickel	+
A 27	Nickel-chromium alloys	+
A 28	Nickel-chromium-iron alloys (without Mo)	+
A 29	Nickel-chromium-molybdenum alloys	+
A 33	Lead and lead alloys	⊕ to −
A 34	Platinum and platinum alloys	+
A 35	Platinum metals (Ir, Os, Pd, Rh, Ru) and their alloys	+
A 36	Tin and tin alloys	+
A 37	Tantalum, niobium and their alloys	+
A 38	Titanium and titanium alloys	+
A 39	Zinc, cadmium and their alloys	+ to −
A 40	Zirconium and zirconium alloys	+
B	**Nonmetallic inorganic materials**	
B 3	Carbon and graphite	+
B 4	Binders for building materials (e.g. concrete, mortar)	⊕ to −
B 5	Acid-resistant building materials and binding agents (putties)	+
B 8	Enamel	+
B 12	Oxide ceramic materials	+
B 13	Metal-ceramic materials (carbides, nitrides)	+
C	**Organic materials/plastics**	
	Thermoplastics	
	Polyolefines and polyvinyl chlorides	+
	Fluorthermoplastics	+
	Fluoropolymers and high-temperature thermoplastics	+ to ⊕
	Polystyrene and copolymere	+
	Polyester	+
	Polyamides	+
	Further thermoplastics	+ to ⊕
	Thermoplastic elastomers	+
	Duroplastics	
	Epoxy resins	+
	Unsaturated polyester resins	+
	Vinyl ester resins	+
	Polyurethanes	+
	Phenolic resins	+
	Furanic resins	+
	Further duroplastics	+ to ⊕
	Elastomers	+ to −
D	**Materials with special properties**	
D 1	Layers, coatings and linings	+
D 2	Seals	+
D 3	Composite materials	+

* Any notes in the text are entered under the same number as the materials, (for example A1, B5, C7)
** + resistant/⊕ fairly resistant/− unsuitable. Where no indication of corrosion resistance is made, experimental data is not available.

Warranty disclaimer

This book has been compiled from literature data with the greatest possible care and attention. The statements made in this chapter only provide general descriptions and information.

Even for the correct selection of materials and correct processing, corrosive attack cannot be excluded in a corrosion system as it may be caused by previously unknown critical conditions and influencing factors or subsequently modified operating conditions.

No guarantee can be given for the chemical stability of the plant or equipment. Therefore, the given information and recommendations do not include any statements, from which warranty claims can be derived with respect to DECHEMA e.V. or its employees or the authors.

The DECHEMA e.V. is liable to the customer, irrespective of the legal grounds, for intentional or grossly negligent damage caused by their legal representatives or vicarious agents.

For a case of slight negligence, liability is limited to the infringement of essential contractual obligations (cardinal obligations). DECHEMA e.V. is not liable in the case of slight negligence for collateral damage or consequential damage as well as for damage that results from interruptions in the operations or delays which may arise from the deployment of this book.

Preliminary remarks
Table of contents

V 1	Introduction	5
V 2	Physical and chemical properties	5
V 3	Production	14
V 3.1	Carbon dioxide gas	14
V 3.2	Liquid carbon dioxide	14
V 3.3	Solid carbon dioxide	14
V 4	Storage and transportation	14
V 5	Applications	15
V 6	Reactions between materials and carbon dioxide	15
V 6.1	Types of damage to metal materials and influencing parameters	16
V 6.1.1	Damage by gaseous carbon dioxide at high temperatures	16
V 6.1.2	Damage by aqueous carbon dioxide solutions	16
V 6.2	Kinds of damage of organic materials and influencing parameters	17
V 7	Corrosion protection	29
V 7.1	Coatings and films	29
V 7.2	Cladding	29
V 7.3	Multilayer structures	29

V 1 Introduction

Carbon dioxide with the chemical symbol CO_2 is a gas at room temperature and the anhydride of carbonic acid H_2CO_3. Pure carbon dioxide is colorless as well as almost odorless and tasteless. Though not being toxic, carbon dioxide does not sustain respiration, and therefore higher concentrations may be lethal (8 to 10 % after 30 to 60 minutes). It is not flammable and can be used as an extinguishing agent. It is about 1.5 times heavier than air, and hence settles to the ground. Carbon dioxide is sold in liquefied form in gray pressurized gas bottles or in solid form as "dry ice".

V 2 Physical and chemical properties

Some of the physical properties are given in Table 1 [1–3]. Since the melting point is higher than the boiling point solid carbon dioxide ("carbonic acid snow", "dry ice") sublimes directly, i.e. under normal pressure and with a sublimation temperature of 194.65 K (–78.5°C) it passes directly from the solid into the gaseous state. Since its critical temperature of 304 K (31°C) is fairly high, carbon dioxide can be easily liquefied under pressure. To this end, pressure of 56.5 bar at room temperature, of 34.4 bar at 273 K (0°C), of 19.3 at 253 K (–20°C) and of 6.6 bar at 223 K (–50°C) is required [4].

Figure 1 shows the thermal conductivity of liquid carbon dioxide as a function of temperature for several pressures [2].

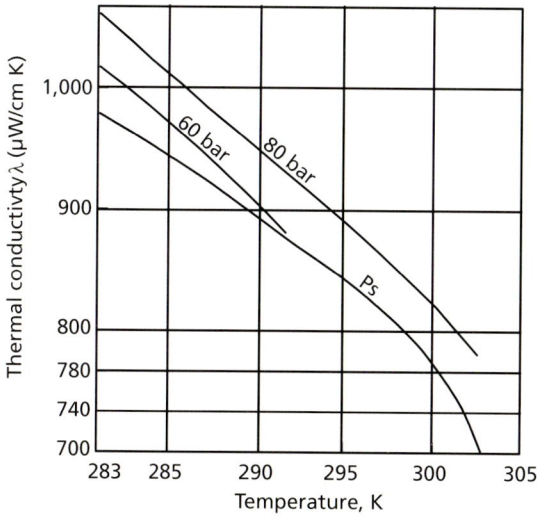

Figure 1: Thermal conductivity of liquid carbon dioxide as a function of temperature and pressure [2]
P_S: Pressure at boiling point

Figure 2 depicts the temperature dependence of the specific heat of carbon dioxide [2].

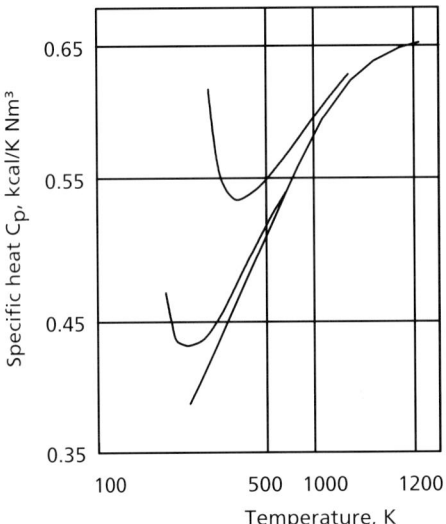

Figure 2: Relationship between specific heat and temperature of carbon dioxide at various pressures [2].

To a major extent, CO_2 is soluble in water and the aqueous solution is acidic since a small portion (about 0.1 %) of the dissolved CO_2 formally reacts with water according to Equation 1, forming carbonic acid.

Equation 1 $CO_2 + H_2O \rightarrow H_2CO_3$

Although carbonic acid having a dissociation constant of $K_1 = 1.3 \times 10^{-4}$ is a medium acid, the entire solution acts like a weak acid only since about 99.9 % of the dissolved CO_2 do not react with water. An aqueous solution saturated with carbon dioxide at normal pressure has a pH value of 3.7 [2].

Being a dibasic acid, carbonic acid forms two types of salts:
hydrogen carbonates (bicarbonates) $MeHCO_3$
carbonates Me_2CO_3
(Me = monovalent metal ion)

The carbonates with the exception of alkali carbonates are hardly soluble in water, whereas all hydrogen carbonates, except sodium hydrogen carbonate ($NaHCO_3$) are easily soluble in water. If the temperature is increased, hydrogen carbonates split off CO_2 and form carbonates according to Equation 2.

Equation 2 $2\ MeHCO_3 \leftrightarrow Me_2CO_3 + CO_2 + H_2O$

Reversely, hydrogen carbonates are formed when CO_2 is passed into aqueous carbonate solutions as described in Equation 3 using the example of an alkali and alkaline earth carbonate.

Equation 3 $\quad Na_2CO_3 + CO_2 + H_2O \rightarrow 2\ NaHCO_3$
$\qquad\qquad\qquad CaCO_3 + CO_2 + H_2O \rightarrow Ca(HCO_3)_2$

These reactions are of relevance to the formation of cover layers on metals in aqueous carbonic acid solutions.

The solubility of carbon dioxide in water decreases as the temperature increases (Figure 3), it increases (Table 2) with pressure.

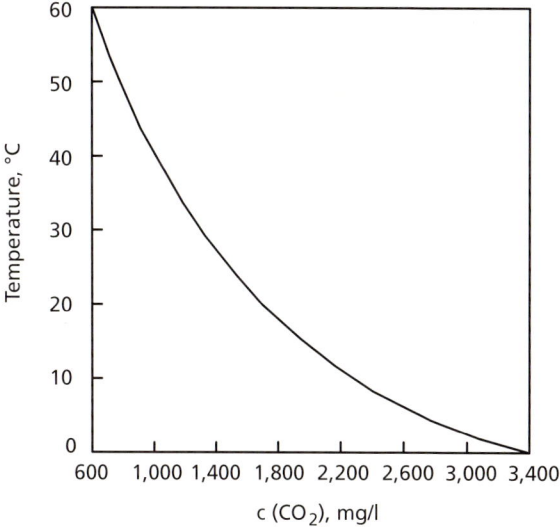

Figure 3: Solubility of carbon dioxide in water at a carbon dioxide pressure of 1 bar as a function of temperature [5]

The amount of a gas dissolved in a liquid is proportional to its partial pressure over the liquid. The following holds true

$$\text{dissolved amount of gas } \frac{Ncm^3 \cdot gas}{g \cdot liquid} = \lambda p_{gas}$$

λ is the technical solubility coefficient depending on temperature. In Table 3 the value λ for carbon dioxide in water is indicated for various temperatures. The values in the table correspond to the amount of carbon dioxide in Ncm^3 (cm^3 at 273 K (0°C), 1 bar) dissolved at saturation of 1 g of water at the relevant temperature if the partial pressure of the carbon dioxide is 1 bar.

Parameter	Unit	Value
Molar mass	g/mol	44.01
Molar volume	l	22.26
Specific weight (weight per liter)	g/l	1.977
Melting point (at 5.2 bar)	K (°C)	216.55 (−56.6)
Boiling point (sublimation temperature)	K (°C)	194.65 (−78.5)
Density in the liquid state (293 K (20°C); 56.3 bar)	g/cm^3	0.766
Density, gaseous (288 K (15°C), 1.013 bar)	kg/m^3	1.85
Density, gaseous (273 K (0°C), 1.013 bar)	kg/m^3	1.98
Relative density, gaseous (air = 1)		1.5291
Vapor pressure at 293 K (20°C)	bar	58.5
Solubility in water (293 K (20°C), 1.013 bar)	g/l	1.7
Heat of evaporation (sublimation heat)	kJ/kg	573.02
Specific heat (293 K (20°C), 1.013 bar)	J/(mol K)	0.9225
Thermal conductivity (288 K (15°C) and 1.013 bar)	W/(cm K)	$1.64 \cdot 10^{-14}$
Critical temperature	K (°C)	304.19 (31.04)
Critical pressure	bar	73.83
Critical density	kg/m^3	468

Table 1: Physical properties of carbon dioxide [1]

Pressure in bar	1	2	3	4
Dissolved liters of CO_2	1	2	3	4

Table 2: Solubility of carbon dioxide in 1 liter of water at room temperature and various pressures [1]

K (°C)	273 (0)	278 (5)	283 (10)	288 (15)	293 (20)	298 (25)	303 (30)	313 (40)	323 (50)	333 (60)	343 (70)
λ	1.658	1.378	1.159	0.987	0.851	0.738	0.646	0.516	0.423	0.353	0.30

Table 3: Technical solubility coefficient λ of carbon dioxide in water for 1 bar and various temperatures [2]

Since the amount of the dissolved CO_2 and, therefore, also the amount of the carbonic acid formed increase as the pressure increases, the pH value of the aqueous solution decreases as shown in Figure 4 and Figure 5.

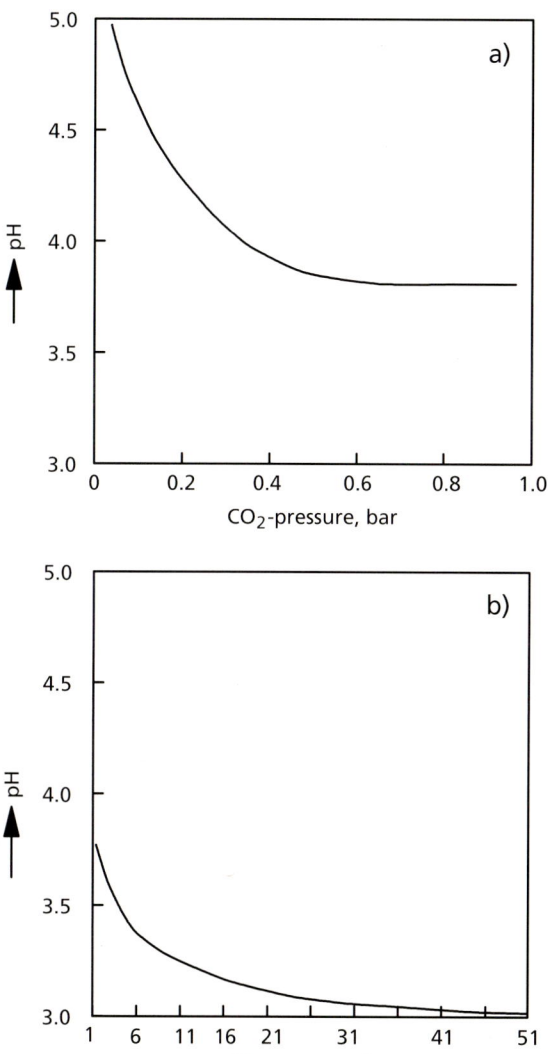

Figure 4: Dependence of the pH value of an aqueous solution on the partial pressure of CO_2 (T = 298 K (25°C)) [79]

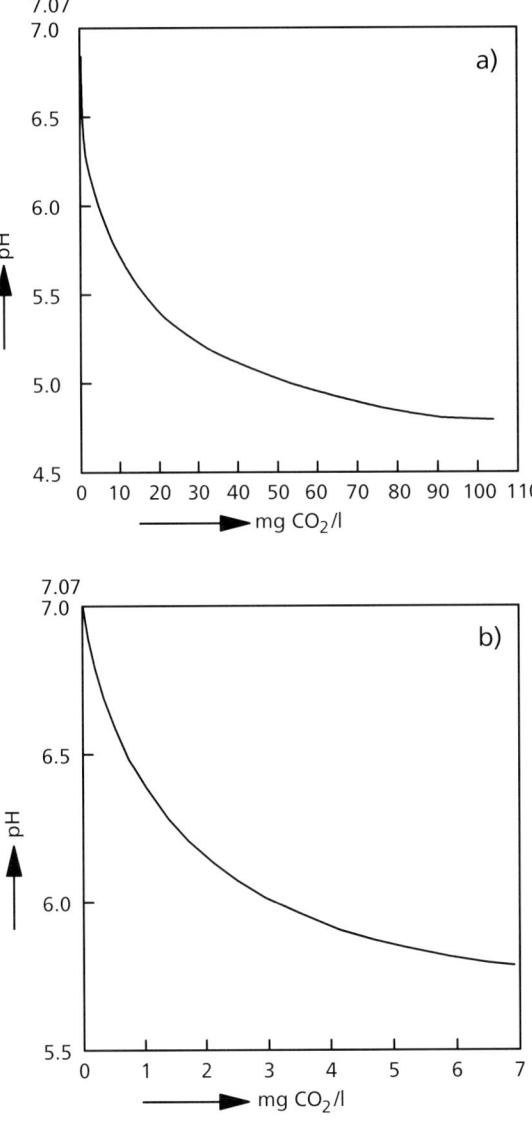

Figure 5: Dependence of an aqueous CO_2 solution on the dissolved amount of CO_2 (T = 298 K (25°C)) [90]

In all natural waters containing dissolved carbonates and hydrogen carbonates as well as dissolved carbon dioxide, an equilibrium exists between carbonic acid, its salts and the dissolved carbon dioxide [6]. This process is described below by using the scheme in Figure 6.

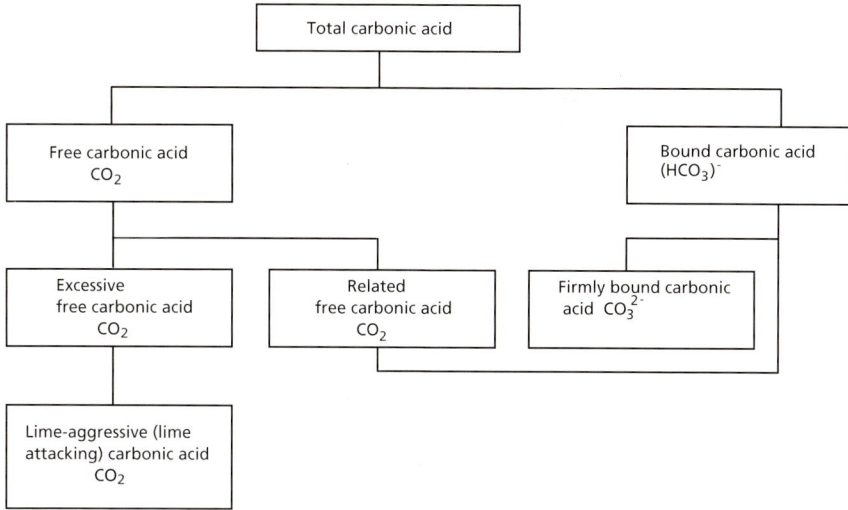

Figure 6: Possible types of "carbonic acid" in aqueous solutions [7]

The terms here are used with the following meanings:

Total carbonic acid refers to the sum of dissolved carbon dioxide and the carbonic acid formed.

$$[CO_{2total}] \; \Sigma \; [CO_{2free}] \; + \; [HCO_3^-] \; + \; [CO_3^{2-}]$$

Free carbonic acid includes both the dissolved gaseous carbon dioxide and the smaller amount (0.1 %) of carbonic acid (H_2CO_3) and its dissociated anions $[CO_3^{2-}]$ and $[HCO_3^-]$.

$$[CO_{2free}] \; \Sigma \; [CO_{2dissolved}] \; + \; [H_2CO_3]$$

Bound carbonic acid is the share of carbon dioxide bound as hydrogen carbonate mainly to calcium and/or magnesium in natural water. The hydrogen carbonates are dissociated and bring about the carbonate hardness of the water.

Firmly bound carbonic acid exists in the form of carbonates.

Related free carbonic acid refers to that portion of dissolved carbon dioxide, which is necessary to keep the hydrogen carbonates in solution.

Excessive free carbonic acid (carbonic acid excess) is the share of carbon dioxide existing in excess of that. It is the aggressive form responsible for corrosion reactions of metals.

Lime-aggressive (lime-attacking) carbonic acid is that portion of the excessive carbonic acid, which reacts with insoluble alkaline earth carbonates according to Equation 3 and can dissolve them to form hydrogen carbonates.

Often, the **lime-carbonic acid balance** of water is used to evaluate the aggressiveness of carbonated waters. The lime-carbonic acid balance is defined as the state when the water, on the one hand, does not precipitate lime however, on the other hand, does not exhibit any corrosive properties. But water with a lime-carbonic acid balance causes the formation of a protective layer on metals only if its oxygen content and the carbonate hardness are sufficient.

The waters existing in nature contain a number of salts in addition to the dissolved gases (oxygen, nitrogen, carbon dioxide). The salts of calcium and magnesium (chlorides, sulfates and hydrogen carbonates) are called hardeners. When heated, the soluble hydrogen carbonates of the alkaline earths are converted into insoluble carbonates and precipitate as "scale" or "limestone". This part of the water hardness, which can be removed by heating, is referred to as carbonate hardness (temporary hardness) in contrast to non-carbonate hardness (lasting or permanent hardness). It results from the sulfate and chloride ions the alkaline earth salts of which cannot be precipitated by heating. The total hardness of the water is the sum of carbonate hardness and non-carbonate hardness.

The lime-carbonic acid balance over the carbonate hardness in Figure 7 facilitates evaluation of the water aggressiveness. All waters beneath the curve contain related carbonic acid and hence are not aggressive to lime or metal. They tend to precipitate calcium carbonate and may cause the formation of a protecting cover layer. Waters above the equilibrium curve contain excessive free and, therefore, aggressive carbonic acid.

The water hardness is usually referred to as the water's degree of hardness with ranges and values not uniformly defined on an international scale. In Germany one degree of German hardness (1°dH or 1°d) was formerly defined as a content of 10.00 mg/l CaO + MgO. Later the molar equivalent unit millival per liter (mval/l)

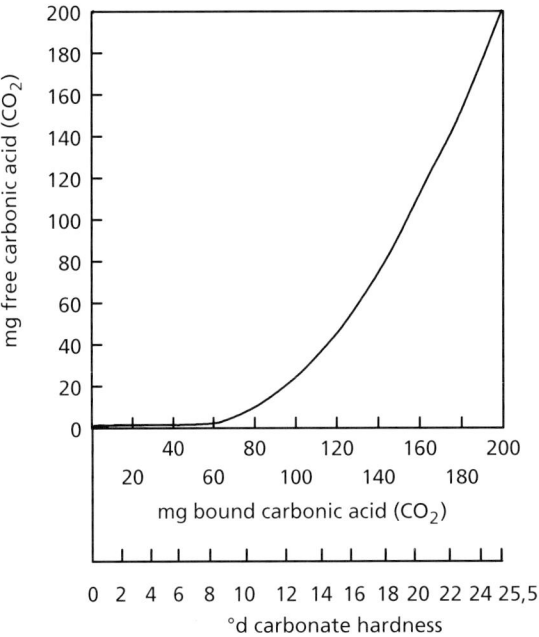

Figure 7: Connection between free carbonic acid (above the curve) and bound carbonic acid (beneath the curve) [8]

was used to indicate the hardness of water. Today, only molar units according to the International System of Units are officially permitted and the content of alkaline earth ions is indicated as the total water hardness in mols per liter or in millimols per liter (mmol/l) given the low concentrations. According to the new system 1°dH corresponds to a content of alkaline earth ions of 5.60 mmol/l.

According to the common classification waters with hardness values of

< 7°dH	are considered soft
from 7°d to 14°dH	medium-hard
from 14°d to 21°dH	hard
> 21°dH	very hard

In the new German Washing and Cleansing Agents Act *(Wasch- und Reinigungsmittelgesetz – WRMG)* as effective from May 2007 the hardness ranges were adjusted to European standards, replacing the unit millimols total hardness per liter by the unit millimols calcium carbonate per liter. Table 4 shows the hardness ranges as defined in the WRMG.

Hardness range	mmol $CaCO_3$/l	°dH
Soft	less than 1.5	less than 8.4
Average	1.5 to 2.5	8.4 to 14
Hard	more than 2.5	more than 14

Table 4: Classification of hardness ranges according to the German Washing and Cleansing Agents Act

Table 5 depicts several common international hardness values with the relating conversion factors [9].

	Alkaline earth ions mmol/l	Alkaline earth ions mval/l	German hardness °d	English hardness °e	French hardness °f	ppm $CaCO_3$ °US
Alkaline earth ions mmol	1.00	2.00	5.60	7.02	10.00	100.0
Alkaline earth ions mval/l	0.50	1.00	2.80	3.51	5.00	50.00
German hardness °d	0.18	0.357	1.00	1.25	1.78	17.90
English hardness °e	0.14	0.285	0.798	1.00	1.43	14.30
French hardness °f	0.10	0.200	0.560	0.702	1.00	10.00
ppm $CaCO_3$ °US	0.01	0.020	0.056	0.0702	0.10	1.00

Table 5: Conversion of common hardness value [9]

V 3 Production

V 3.1 Carbon dioxide gas

A major part of the world production of carbon dioxide is obtained as a byproduct during the hydrogen generation from natural gas for ammonia synthesis according to the reaction of Equation 4.

Equation 4 $\quad CH_4 + 2\,H_2O \rightarrow CO_2 + 4\,H_2$

In addition, carbon dioxide is obtained from flue gas purification or may be obtained from calcium carbonate in a reversible reaction

Equation 5 $\quad K_2CO_3 + H_2O + CO_2 \leftrightarrow 2\,KHCO_3$

or from the ethanolamine process according to

Equation 6 $\quad 2\,HOC_2H_4NH_2 + H_2O + CO_2 \leftrightarrow (HOC_2H_4NH_3)_2CO_3$

[1].

V 3.2 Liquid carbon dioxide

As the values of the critical temperature (304 K (31°C)) and the critical pressure (7.4 MPa) of CO_2 show, CO_2 can be liquefied at any temperature between 304 K (31°C) and the triple point (216.55 K (−56.6°C)) by applying the necessary pressure and removing the compression heat. To this end, a pressure of 5.6 MPa is sufficient at 293 K (20°C).

V 3.3 Solid carbon dioxide

Solid carbon dioxide (dry ice) is obtained if liquid carbon dioxide is relieved at atmospheric pressure. Due to the required evaporation energy the temperature is reduced to a level below the sublimation value such that solid carbonic acid snow is generated. This snow is pressed to form solid blocks sold as "dry ice".

V 4 Storage and transportation

Carbon dioxide is stored in its liquid form under pressure in storage vessels of steel and is transported in gray steel bottles.

V 5 Applications

About 50 % of the produced carbon dioxide is used for the production of chemical products, such as methanol and urea. Other applications include its use as an inert shielding gas, heat carrier or cooling agent.

A special application of carbon dioxide is its use in petroleum production to improve the yield of aging oil wells. A driving force is required to bring the crude oil to the surface. In newly tapped oil wells, this is effected by the gases dissolved in the oil. With increasing age of the well other or additional sources of energy are required. Frequently, water is used in the first instance to which, finally, a soluble gas, such as CO_2, is added to reduce the interfacial tension of the encapsulated oil drops and thus to mobilize the remaining reserves. Since the pH value of the press water decreases as a result of the high share of dissolved CO_2 at the prevailing pressures, such waters become strongly aggressive, in particular together with hydrogen sulfide dissolved at the same time, and may cause considerable corrosion damage to the materials of oil extracting facilities.

V 6 Reactions between materials and carbon dioxide

At ambient temperatures the gaseous carbon dioxide is very unreactive since the CO_2 molecule is very stable and needs a high temperature to split into carbon monoxide and oxygen according to Equation 7.

Equation 7 $2\,CO_2 \leftrightarrow 2\,CO + O_2$

This reaction can be used to produce gas atmospheres with low oxygen activity [10]. So, carbon dioxide acts as an oxidant at high temperatures and leads to the formation of metal oxides when reacting with metals at temperatures above about 773 K (500°C). This also applies to mixtures of CO_2 and CO as well as to mixtures of CO_2 and H_2O.

It is not the absolute pressure of CO_2 and CO but their ratio, which determines the partial pressure of oxygen

Equation 8 $p_{O_2} = \dfrac{K_1 (p_{CO_2})^2}{p_{CO}}$

However, mixtures of CO_2 and CO cannot only exert an oxidizing but also a carbonating effect according to Equation 9.

Equation 9 $2\,CO \leftrightarrow CO_2 + C$

The carbon activity can be calculated according to

Equation 10 $a_c = \dfrac{K_2 (p_{CO})^2}{p_{CO_2}}$

Depending on the amount of the dissolved carbon dioxide, aqueous solutions of CO_2 reach pH values of 3.5 and, hence, may attack metals as a weak acid (refer to Figure 4).

V 6.1 Types of damage to metal materials and influencing parameters

V 6.1.1 Damage by gaseous carbon dioxide at high temperatures

Oxidation by oxygen formed according to Equation 7 leads to the formation of metal oxides in metals at temperatures above about 773 K (500°C).

During **carburization** in hot gases with low carbon activity the carbon formed according to Equation 9 from the gas atmosphere diffuses into the metal matrix and leads to the formation of carbides of the M_7C_3 and $M_{23}C_6$ types (with M = ion, chromium, nickel, e.g. in FeCrNi alloys). First the carbide $M_{23}C_6$ is formed, while precipitation into the inside of the material is progressing. Then, this carbide is converted into M_7C_3. This internal carbide formation influences the mechanical properties of the material and, in particular, reduces its ductility such that cracks may occur.

In hot gases with high carbon activities ($a_c > 1$), i.e. with a high CO content, a special form of catastrophic carburization can be observed at temperatures in a range from about 673 to 1073 K (400°C to 800°C), known as "**metal dusting**". During this process the unstable carbide M_3C is formed on the material surface and at the grain boundaries, which decomposes into $M_3C \rightarrow 3\,M + C$, the material being converted into fine powder of metal particles and graphite. All metals and alloys, which are able to dissolve carbon, are sensitive to this type of damage, e.g. ion, nickel, cobalt and their alloys. Carbon is absorbed by the metal from the atmosphere up to a stage of oversaturation, leading to the formation of graphite and destroying the material. For detailed information regarding the phenomenon of metal dusting reference is made to [11].

V 6.1.2 Damage by aqueous carbon dioxide solutions

Aqueous solutions of CO_2 may attack metals as weak acids. Carbonic acid formed in aqueous carbonic acid solutions according to Equation 1 can react with metals according to

Equation 11: $\quad Me(I) + H_2CO_3 \rightarrow MeHCO_3 + H$
$\qquad\qquad\qquad Me(II) + H_2CO_3 \rightarrow MeCO_3 + 2\,H$,

releasing atomic hydrogen. On the one hand, this may cause a more or less uniform surface corrosion on materials, which are not sufficiently resistant in weak acids, and on the other hand, the hydrogen released during this process may also cause hydrogen embrittlement or hydrogen-induced stress corrosion cracking in sensitive materials, e.g. in higher strength steels.

V 6.2 Kinds of damage of organic materials and influencing parameters

The term of corrosion of organic polymer materials covers a range of various forms of damage and adverse changes, which may cause a reduced operating life of the materials. [12] provides a systematization of the corrosion of organic materials based on its manifestations (e.g. uniform material removal, pitting, formation of scars and cracks) or on the basis of the corrosion mechanism (e.g. chemical or microbiologic attack, physically induced changes as a result of sorption of low molecular weight substances, extraction or migration of components, effects of mechanical stresses) [12].

Regarding the action of carbonic acid or aqueous solutions of carbon dioxide, the following types of damage need to be considered, in particular:

- chemical attack on the material resulting from hydrolysis,
- sorption of carbon dioxide and water and the resulting swelling of the material,
- dissolving out of water-soluble components from the material,
- permeation of carbon dioxide and water through the material and thus reduction or loss of its blocking or insulating effect.

Usually, the intensity of damage increases as the temperature, the exposure and the stresses or strains increase, which exist in the material or the component or are applied from outside. Moreover, aging of a material may be caused by atmospheric influences, such as light, changing temperatures, oxygen, humidity and other factors, fostering the subsequent attack by carbonic acid.

Chemical attack by hydrolysis mainly affects polymers with hydrolysable groups, first of all ester groups, but may also affect materials with fillers or fibers subjected to a hydrolytic attack. Compared to caustic soda lye or potash, the hydrolytic attack by carbonic acid, however, is much weaker.

The reaction products of hydrolysis of the polymer or other constituents of the material diffuse out of the material into the medium, leading to a material consumption of the material. However, the reaction products can also remain in the material and, for example, collect in small cavities of the material where they then act as seeds for osmotic blisters.

The absorption of carbon dioxide and water as a result of sorption and diffusion and the resulting swelling of the polymer mainly depend on the temperature, the chemical structure of the polymer and the degree of cross-linking. Normally, rising temperatures and a decreasing degree of cross-linking lead to higher absorption of the fluid. Polar groups such as OH or CONH groups increase above all the amount of water take up. The fluid absorption causes noticeable changes in physical properties, e.g. a reduction of strength, stiffness, hardness and glass transition temperature; moreover, also dimensional changes of the components need to be considered.

Regarding organic materials, which are used for corrosion protection purposes in the form of coatings and linings and are employed in the manufacture of chemical equipment (refer to Section D), a high barrier effect towards low molecular weight substances is important. Thus, permeation, i.e. the transport of molecules of carbon dioxide and water molecules through a polymer layer with exit on the rear side of the material must be taken into account as well. The damage caused by permeation often occurs behind the material, e.g. in the area of the steel surface in case of coated steel components or, in the case of composite components, at the interface of two plastic materials with different permeability.

As key parameter serving as a measure for the permeability of a material, the permeability coefficient that is independent of the layer thickness, but heavily dependent on temperature, has been defined. Usually the permeability coefficient increases significantly with increasing temperature.

Table 6 lists the permeability coefficients of carbon dioxide of plastic materials.

Remarkably, the permeability coefficients indicated for a specific polymer in the literature may differ widely; reasons may be: different methods of measurement, different pressure differentials chosen for the measurements, variable degree of crystallization of semi-crystalline polymers, different stretching of films, varying monomer contents of copolymers, different modification agents and additives influencing the permeability.

Additional permeation data is indicated under the individual materials.

Polymer	Polymer abbreviation[1]	Temperature K (°C)	Permeability coefficient for carbon dioxide $10^{-13} \times cm^3$ (STP) cm/cm^2 s Pa	Remarks	References
a) Thermoplastics					
Polyethylene, low density	PE-LD	293 (20)	7.5–8.1		[13]
		296 (23)	11.6		[14]
		298 (25)	9.5		[15, 16]
		298 (25)	9.83		[17]
		298 (25)	12.0	CO_2 pressure: 4 bar	[18]
		303 (30)	11.6		[13]
		308 (35)	15.8	crystallinity: 43%	[19]
		313 (40)	22	amorphous phase: 70 Vol%	[20]
		323 (50)	30.1		[13]
		333 (60)	41	amorphous phase: 70 Vol%	[20]
		349 (76)	64	amorphous phase: 70 Vol%	[20]

STP = Standard Temperature and Pressure (273.15 K (0°C); 1.13 × 10^5 Pa)
RT = room temperature
[1] Abbreviation according to [41] and [42]

Table 6: The permeability coefficients of polymers for carbon dioxide

Table 6: Continued

Polymer	Polymer abbreviation[1]	Temperature K (°C)	Permeability coefficient for carbon dioxide $10^{-13} \times cm^3$ (STP) cm/cm^2 s Pa	Remarks	References
Linear polyethylene, low density	PE-LLD	293 (20)	7.5–8.1		[13]
		303 (30)	11.6		[13]
		303 (30)	19.9	film unstretched	[21]
		303 (30)	16.0	film stretched in longitudinal direction	[21]
		303 (30)	16.9	film stretched in transverse direction	[21]
		323 (50)	30.1		[13]
Polyethylene, medium density	PE-MD	312 (39)	8.1	amorphous phase: 53 Vol%	[20]
		333 (60)	16.2	amorphous phase: 53 Vol%	[20]
		334 (61)	17	amorphous phase: 53 Vol%	[20]
		353 (80)	29	amorphous phase: 53 Vol%	[20]
Polyethylene, high density	PE-HD	approx. 293 (20)	3.2		[22]
		298 (25)	3.3		[14]
		298 (25)	3.2	stretch ratio: 1.0	[16]
		298 (25)	1.4	stretch ratio: 6.4	[16]
		298 (25)	0.28	stretch ratio: 11.8	[16]
		298 (25)	0.045	stretch ratio: 17.6	[16]
		313 (40)	5.5	amorphous phase: 37 Vol%	[20]
		333 (60)	9.6	amorphous phase: 37 Vol%	[20]
		354 (81)	19	amorphous phase: 37 Vol%	[20]
Polypropylene	PP	293 (20)	2.3–2.7	undrawn film	[13]
		approx. 293 (20)	2.5		[22]
		298 (25)	2.8		[14]
		303 (30)	3.7–4.2	undrawn film	[13]
		306 (33)	5.43	crystallinity: 43%	[16]
		306 (33)	4.64	crystallinity: 50%	[16]
		306 (33)	4.05	crystallinity: 58%	[16]
		308 (35)	8.9	crystallinity: 50%	[19]
		323 (50)	11.6–13.4	undrawn film	[13]

STP = Standard Temperature and Pressure (273.15 K (0°C); 1.13×10^5 Pa)
RT = room temperature
[1] Abbreviation according to [41] and [42]

Table 6: The permeability coefficients of polymers for carbon dioxide

Table 6: Continued

Polymer	Polymer abbreviation[1]	Temperature K (°C)	Permeability coefficient for carbon dioxide $10^{-13} \times cm^3$ (STP) cm/cm^2 s Pa	Remarks	References
Polyisobutylene	PIB	308 (35)	6.5		[19]
Polyvinyl chloride, without plasticizers	PVC-U	293 (20)	0.093		[14]
		approx. 293 (20)	0.09		[22]
		296 (23)	0.14–0.19		[14]
		RT	0.12–0.17	undrawn film	[13]
		298 (25)	0.12		[16]
		308 (35)	0.138–0.182		[16]
		308 (35)	0.14	crystallinity: about 10 %	[19]
Polyvinyl chloride containing plasticizer	PVC-P	293 (20)	3.9		[14]
Polyvinylidene chloride	PVDC	298 (25)	0.017–0.2		[14]
Polytetrafluorethylene	PTFE	296 (23)	1.6–2.4		[14]
		298 (25)	7.5		[16]
		298 (25)	8.8		[15]
		RT	17.4		[13]
Tetrafluoroethylene-perfluoropropyl-vinylether copolymer	PFA	297 (24)	10.0		[16]
		RT	8.1		[13]
Tetrafluoroethylene-hexafluoropropylene copolymer	FEP	298 (25)	9.6		[15]
		298 (25)	9.53		[16]
		RT	5.4		[13]
Ethylene tetrafluoro-ethylene copolymer	ETFE	296 (23)	1.1		[14]
		296 (23)	1.49		[16]
		296 (23)	1.88	standard type	[23]
		296 (23)	3.45	modified, crack resistant	[23]
		RT	1.5		[13]
Polychlorotrifluoro-ethylene	PCTFE	RT	0.17		[13]
		313 (40)	0.072–0.18		[14]
		313 (40)	0.158	crystallinity: 30 %	[16]
		313 (40)	0.036	crystallinity: 80 %	[16]

STP = Standard Temperature and Pressure (273.15 K (0°C); 1.13×10^5 Pa)
RT = room temperature
[1] Abbreviation according to [41] and [42]

Table 6: The permeability coefficients of polymers for carbon dioxide

Table 6: Continued

Polymer	Polymer abbreviation[1]	Temperature K (°C)	Permeability coefficient for carbon dioxide $10^{-13} \times cm^3$ (STP) cm/cm^2 s Pa	Remarks	References
Ethylene chlorotri-fluoroethylene copolymer	ECTFE	296 (23)	0.457		[16]
		296 (23)	0.49		[14]
		RT	0.46		[13]
Polyvinylidene fluoride	PVDF	approx. 293 (20)	0.24		[22]
		RT	0.12		[13]
		308 (35)	0.39	crystallinity: 54 %	[19]
		321 (48)	1.16	homopolymer	[24]
		343 (70)	4.85	amorphous phase: 53 Vol%	[20]
		374 (101)	13.7	amorphous phase: 53 Vol%	[20]
		383 (110)	16.7	amorphous phase: 53 Vol%	[20]
		405 (132)	30	amorphous phase: 53 Vol%	[20]
Polyvinyl fluoride	PVF	296 (23)	0.049		[14]
		296 (23)	0.069		[16]
		RT	0.069		[13]
		308 (35)	0.20	crystallinity: 38 %	[19]
		308 (35)	0.20–0.388		[16]
Polysulfone	PSU	296 (23)	2.70	CO_2 pressure: 34 bar	[25]
		296 (23)	4.2–4.3		[14, 26]
		308 (35)	4.2		[16]
		313 (40)	4.9	CO_2 pressure: 8 atm	[27]
		313 (40)	3.95	CO_2 pressure: 30 atm	[27]
Polyethersulfone	PESU	294 (21)	1.95	CO_2 pressure: 27 bar	[25]
		296 (23)	3.0–3.6		[26]
Polyaryl ether ether ketone	PEEK	298 (25)	2.03		[16]
		328 (55)	3.40		[16]
		348 (75)	4.54		[16]
Polyetherimide	PEI	294 (21)	0.63	CO_2 pressure: 28 bar	[25]

STP = Standard Temperature and Pressure (273.15 K (0°C); 1.13×10^5 Pa)
RT = room temperature
[1] Abbreviation according to [41] and [42]

Table 6: The permeability coefficients of polymers for carbon dioxide

Table 6: Continued

Polymer	Polymer abbreviation[1]	Temperature K (°C)	Permeability coefficient for carbon dioxide $10^{-13} \times cm^3$ (STP) cm/cm^2 s Pa	Remarks	References
Polyimide	PI	296 (23)	0.20		[14]
		298 (25)	0.00692–1.36	depending on dicarboxylic acid anhydride/diamine	[16]
Polybenzimidazole	PBI	298 (25)	6.9		[14]
Polystyrene	PS	296 (23)	6.0		[14]
		298 (25)	7.88		[17]
		RT	0.87	biaxially oriented film	[13]
		298 (25)	0.46	oriented film	[14]
		298 (25)	7.9	biaxially oriented film	[16]
		298 (25)	6.0	stretch ratio: 1.0	[16]
		298 (25)	2.18	stretch ratio: 3.1	[16]
		298 (25)	0.75	stretch ratio: 5.0	[16]
Polystyrene, shock-resistant	PS-HI	296 (23)	11.6		[14]
Graft copolymer of styrene and acrylonitrile on polybutadiene	ABS	293 (20)	2.23	CO_2 pressure: 2–10 bar	[28]
		303 (30)	2.68	CO_2 pressure: 2–10 bar	[28]
		313 (40)	2.98	CO_2 pressure: 2–10 bar	[28]
		323 (50)	3.71	CO_2 pressure: 2–10 bar	[28]
Graft copolymer of styrene and acrylonitrile on polyacrylester	ASA	293 (20)	6.9–9.3		[14]
Acrylonitrile styrene copolymer Acrylonitrile/	SAN				
styrene 86/14		298 (25)	0.011		[16]
66/34		298 (25)	0.16		[16]
57/43		298 (25)	0.27		[16]
39/61		298 (25)	1.0		[16]

STP = Standard Temperature and Pressure (273.15 K (0°C); 1.13×10^5 Pa)
RT = room temperature
[1] Abbreviation according to [41] and [42]

Table 6: The permeability coefficients of polymers for carbon dioxide

Table 6: Continued

Polymer	Polymer abbreviation[1]	Temperature K (°C)	Permeability coefficient for carbon dioxide $10^{-13} \times cm^3$ (STP) cm/cm^2 s Pa	Remarks	References
Methacrylonitrile styrene copolymer Methacrylonitrile/					
styrene 97/3		298 (25)	0.0059		[16]
82/18		298 (25)	0.038		[16]
61/39		298 (25)	0.21		[16]
38/62		298 (25)	0.88		[16]
18/82		298 (25)	2.4		[16]
Copolymer of styrene and styrene sulfonic acid a) Sodium polysalt					
0 mol% sulfonic acid		296 (23)	10.58		[16]
15.2 mol% sulfonic acid		296 (23)	3.53		[16]
27.5 mol% sulfonic acid		296 (23)	2.18		[16]
b) Magnesium polysalt					
15.2 mol% sulfonic acid		296 (23)	2.25		[16]
27.5 mol% sulfonic acid		296 (23)	1.65		[16]
Polyacrylonitrile	PAN	298 (25)	0.00060		[16]
Polyethylene terephthalate	PET	296 (23)	0.058–0.098	oriented film	[14]
		296 (23)	0.069		[13]
		296 (23)	0.065–0.10		[14]
		298 (25)	0.110	cast film, amorphous	[29]
		298 (25)	0.130–0.134	cast film, biaxially oriented	[29]
		298 (25)	0.227	amorphous	[16]
		298 (25)	0.118	crystallinity: 40%	[16]
		303 (30)	0.081		[13]
		363 (90)	1.0	crystallinity: 40%	[16]
Polybutylene terephthalate	PBT	298 (25)	0.217	amorphous	[16]
Polyethylen naphthalate		298 (25)	0.135–0.165	cast film, biaxially oriented	[29]

STP = Standard Temperature and Pressure (273.15 K (0°C); 1.13×10^5 Pa)
RT = room temperature
[1] Abbreviation according to [41] and [42]

Table 6: The permeability coefficients of polymers for carbon dioxide

Table 6: Continued

Polymer	Polymer abbreviation[1]	Temperature K (°C)	Permeability coefficient for carbon dioxide $10^{-13} \times cm^3$ (STP) cm/cm^2 s Pa	Remarks	References
Polycarbonate	PC	296 (23)	4.2		[14]
		298 (25)	3.53	CO_2 pressure: 31 bar	[25]
		308 (35)	4.5		[16, 30]
Tetramethyl bisphenol-A polycarbonate		298 (25)	9.75	CO_2 pressure: 13 bar	[25]
		308 (35)	13.2		[30]
Poly(2,2,4,4-tetramethyl-cyclobutane carbonate)		308 (35)	57.6	pressed film	[30]
		308 (35)	41.1	cast film of CH_2Cl_2	[30]
		308 (35)	23.1	cast film of $CHCl_3$	[30]
Polytetrabutylene carbonate		308 (35)	3.17		[16]
Polytetramethylene carbonate		308 (35)	13.2		[16]
Polyamide 6	PA6	293 (20)	0.066		[16]
		293 (20)	0.046–0.069		[31]
		296 (23)	0.093–0.14		[14]
		298 (25)	0.056		[17]
Polyamide 66	PA66	278 (5)	0.018	undrawn	[16]
		278 (5)	0.023	stretched	[16]
		296 (23)	0.093–0.14		[14]
		298 (25)	0.052	undrawn	[16]
		298 (25)	0.071	stretched	[16]
Polyamide 11	PA11	293 (20)	0.6–1.2		[32]
		296 (23)	0.007–0.015		[14]
		313 (40)	0.754		[16]
		344 (71)	2.31	amorphous phase: 79 Vol%	[20]
		374 (101)	5.45	amorphous phase: 79 Vol%	[20]
		394 (121)	10	amorphous phase: 79 Vol%	[20]
		404 (131)	12	amorphous phase: 79 Vol%	[20]

STP = Standard Temperature and Pressure (273.15 K (0°C); 1.13×10^5 Pa)
RT = room temperature
[1] Abbreviation according to [41] and [42]

Table 6: The permeability coefficients of polymers for carbon dioxide

Table 6: Continued

Polymer	Polymer abbreviation[1]	Temperature K (°C)	Permeability coefficient for carbon dioxide $10^{-13} \times cm^3$ (STP) cm/cm^2 s Pa	Remarks	References
Polyamide 11 with plasticizer (N-butyl-benzenesulfonamide)	PA11-P	393 (120)	10.6	0 Ma% plasticizer	[20]
		393 (120)	23.5	7.5 Ma% plasticizer	[20]
		393 (120)	36.5	12.5 Ma% plasticizer	[20]
		393 (120)	35.4	19 Ma% plasticizer	[20]
		393 (120)	52.7	29.5 Ma% plasticizer	[20]
Polyamide 12	PA12	296 (23)	0.007–0.015		[14]
Polymethylmethacrylate	PMMA	308 (35)	2.33		[16]
Polyethylmethacrylate		298 (25)	3.79		[16]
		298 (25)	9.8–13.5	CO_2 pressure: 6.9 bar	[33]
		308 (35)	5.26	CO_2 pressure: 1 atm	[34]
		308 (35)	16.1	CO_2 pressure: 25 atm	[34]
		318 (45)	14.3	CO_2 pressure: 6.9 bar	[33]
		326 (53)	18.8	CO_2 pressure: 6.9 bar	[33]
Polyvinyl alcohol	PVAL	298 (25)	0.00924	relative humidity: 0%	[16]
Polyvinylbenzoate		298 (25)	4.17		[16]
		308 (35)	4.74		[16]
		343 (70)	8.06		[16]
Poly(vinyl-p-isopropyl-benzoate)		298 (25)	15.7		[16]
		353 (80)	32.9		[16]
Poly(vinyl-m-methyl-benzoate)		308 (35)	5.05		[16]
Ethylene-vinyl acetate copolymer Vinyl acetate content of copolymer:	EVAC				
19 Ma%		298 (25)	42.8	CO_2 pressure: 4 bar	[18]
33 Ma%		298 (25)	22.5	CO_2 pressure: 4 bar	[18]
50 Ma%		298 (25)	52.5	CO_2 pressure: 4 bar	[18]
70 Ma%		298 (25)	22.5	CO_2 pressure: 4 bar	[18]

STP = Standard Temperature and Pressure (273.15 K (0°C); 1.13×10^5 Pa)
RT = room temperature
[1] Abbreviation according to [41] and [42]

Table 6: The permeability coefficients of polymers for carbon dioxide

Table 6: Continued

Polymer	Polymer abbreviation[1]	Temperature K (°C)	Permeability coefficient for carbon dioxide $10^{-13} \times cm^3$ (STP) cm/cm² s Pa	Remarks	References
Polyoxymethylene	POM	296 (23)	0.44		[14]
		296 (23)	1.28		[16]
		298 (25)	1.35		[16]
Poly(oxy-2,6-dimethyl-1,4-phenylene) (Polyphenylene ether)	PPE	298 (25)	56.8		[16]
		298 (25)	61.5	CO_2 pressure: 14 bar	[25]
Cellulose hydrate		298 (25)	0.00353	relative humidity: 0 %	[16]
Ethyl cellulose	EC	298 (25)	84.8		[16]
Cellulose acetate	CA	298 (25)	4.1		[14]
		300 (27)	4.5	CO_2 pressure: 11 bar	[25]
		303 (30)	17.3	with plasticizer	[16]
		308 (35)	4.47		[16]
		308 (35)	2.3	CO_2 pressure: 10 atm	[35]
		308 (35)	4.1	CO_2 pressure: 27 atm	[35]
		308 (35)	6.4	CO_2 pressure: 42 atm	[35]
Cellulose triacetate	CTA	297 (24)	5.5	CO_2 pressure: 10 bar	[25]
Cellulose acetobutyrate	CAB	298 (25)	27.2		[14]
Cellulose nitrate	CN	298 (25)	1.59		[16]
b) Thermoplastic elastomers					
Thermoplastic polyurethane elastomers a) Polyesterol basis	TPE-U	293 (20)	20	Shore A hardness: 80	[36]
		293 (20)	15	Shore A hardness: 85	[36]
		293 (20)	4	Shore A hardness: 90	[36]
		293 (20)	2	Shore A hardness: 95	[36]
b) Polyesterol basis		293 (20)	23	Shore A hardness: 80	[36]
		293 (20)	18	Shore A hardness: 85	[36]
		293 (20)	13	Shore A hardness: 90	[36]
		293 (20)	9	Shore A hardness: 95	[36]

STP = Standard Temperature and Pressure (273.15 K (0°C); 1.13×10^5 Pa)
RT = room temperature
[1] Abbreviation according to [41] and [42]

Table 6: The permeability coefficients of polymers for carbon dioxide

Table 6: Continued

Polymer	Polymer abbreviation[1]	Temperature K (°C)	Permeability coefficient for carbon dioxide $10^{-13} \times cm^3$ (STP) cm/cm^2 s Pa	Remarks	References
c) Rubbers, elastomers					
Natural rubber	NR	298 (25)	115		[16]
Polybutadiene	BR	298 (25)	104		[16]
Polydimethyl butadiene (Methyl rubber)		298 (25)	5.63		[16]
Polybutadiene, hydrated		298 (25)	36.3	crystallinity: 29 %	[16]
Polyisoprene (Isoprene rubber)	IR	301 (28)	128		[37]
Polyoctenamer		301 (28)	75		[37]
Chloroprene rubber	CR	298 (25)	19.2		[16]
Acrylonitrile butadiene rubber	NBR	298 (25)	47.6	acrylonitrile content: 20 %	[16]
		298 (25)	23.2	acrylonitrile content: 27 %	[16]
		298 (25)	13.9	acrylonitrile content: 32 %	[16]
		298 (25)	5.59	acrylonitrile content: 39 %	[16]
		353 (80)	47.2		[38]
Hydrated acrylonitrile-butadiene rubber	HNBR	293 (20)	12	CO_2 pressure: 30 bar	[39]
Isobutene isoprene rubber (Butyl rubber)	IIR	298 (25)	3.89		[16]
Ethylene propylene diene rubber	EPDM	293 (20)	28	CO_2 pressure: 30 bar	[39]
Fluorinated rubber	FKM	293 (20)	15	CO_2 pressure: 20 bar	[39]
		293 (20)	93	CO_2 pressure: 40 bar	[39]
		293 (20)	243	CO_2 pressure: 50 bar	[39]

STP = Standard Temperature and Pressure (273.15 K (0°C); 1.13×10^5 Pa)
RT = room temperature
[1] Abbreviation according to [41] and [42]

Table 6: The permeability coefficients of polymers for carbon dioxide

Table 6: Continued

Polymer	Polymer abbreviation[1]	Temperature K (°C)	Permeability coefficient for carbon dioxide $10^{-13} \times cm^3$ (STP) cm/cm^2 s Pa	Remarks	References
Polyurethane	AU, EU	301 (28)	12.0		[37]
Polydimethylsiloxane	MQ	298 (25)	2430	membrane thickness: 0.17 mm	[17]
		298 (25)	2108	membrane thickness: 1.11 mm	[17]
		301 (28)	975		[37]
		308 (35)	3489		[16]
Polymethylethyl siloxane		308 (35)	1130		[16]
Polymethylpropyl siloxane		308 (35)	1727		[16]
Polymethyloctyl siloxane		308 (35)	695		[16]
Polymethylphenyl siloxane	PMQ	308 (35)	179		[16]
Polydicyclohexyl siloxane		308 (35)	1095	CO_2 pressure: 4.54 atm	[40]
		308 (35)	1140	CO_2 pressure: 7.73 atm	[40]
		308 (35)	1170	CO_2 pressure: 11.1 atm	[40]
		308 (35)	1200	CO_2 pressure: 14.5 atm	[40]

STP = Standard Temperature and Pressure (273.15 K (0°C); 1.13×10^5 Pa)
RT = room temperature
[1] Abbreviation according to [41] and [42]

Table 6: The permeability coefficients of polymers for carbon dioxide

Various test procedures can be used to test the resistance of organic materials to aqueous carbon dioxide solutions and carbon dioxide gas:

- Immersion tests,
- Immersion tests under mechanical stress,
- Creep-depending-on-time tests under internal compression for pipes,
- Tests with single side media stress for coatings and GFRP materials.

Details regarding the individual tests, including applicable standards, are contained in the DECHEMA Corrosion Handbooks "Potassium Hydroxide", "Ammonia" and "Sodium Chloride".

V 7 Corrosion protection

V 7.1 Coatings and films

Generally, corrosion protection can be provided by separating the sensitive material from the relevant corrosive medium by means of a protective film or coating. For dried gaseous carbon dioxide corrosion protection is not required at lower temperatures, and is not possible at high temperatures using organic coatings.

In carbonic acid solutions corrosion protection can be provided by suitable organic coatings or metal coatings.

V 7.2 Cladding

A resistant material can be applied to the carrier material by:
- Roll cladding
- Explosive cladding
- Deposition welding
- Thermal spraying.

Roll cladded or explosively cladded blanks, e.g. sheets or bottoms, are jointed to form the desired unit. Deposition welding or thermal spraying may be also used to protect finished units.

V 7.3 Multilayer structures

For multilayer structures an internal core tube of a resistant material is covered with one or more layers from which the pressure bearing part is manufactured.

A
Metallic materials

A 1 Silver and silver alloys

Even in the presence of oxygen only trace amounts of silver are dissolved in aqueous carbon dioxide solutions (carbonated water). The behavior of silver-rich silver solders is similar. Only little has been reported in the literature about the corrosion rates of silver in carbonated water. One of these publications refers to a very low corrosion of 0 to 0.08 mm/a (0 to 3.15 mpy) [43].

A 2 Aluminium

Due to their passivated layer aluminium materials have a good corrosion resistance in almost neutral or low-chloride aqueous media. As can be seen in Figure 8, the passivated area of aluminium ranges from about pH 4 to pH 8.5 [44].

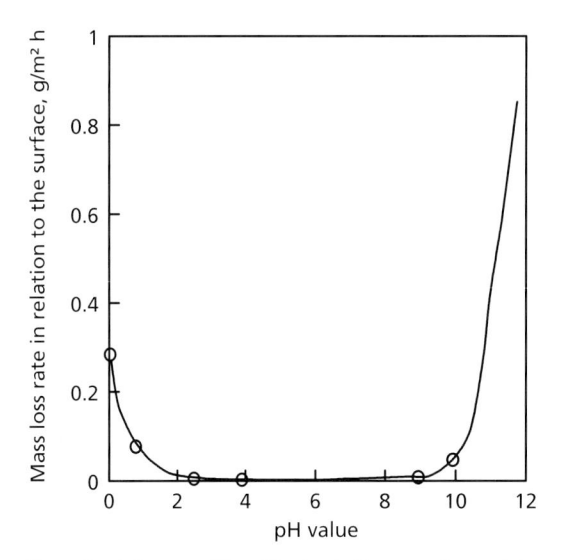

Figure 8: Influence of the pH value to the corrosion resistance of aluminium [44]

Since pH values below 4 in CO_2 containing waters can be reached only at high partial CO_2 pressures, a sufficient resistance of aluminium materials can be expected under normal conditions in such waters.

The studies described in [45] investigated the influence of carbon dioxide on the behavior of pure aluminium Al99,5 in a 3.5% sodium chloride solution, involving a pitting corrosion risk for aluminium as in all chloride ion-containing solutions. The test solutions were either flushed with high-purity carbon dioxide, nitrogen, a mixture of nitrogen with 1% CO_2 or with air, partially contained sodium hydrogen carbonate as a buffering substance. Solutions flushed with pure nitrogen were adjusted to various pH values by adding low amounts of hydrochloric acid or NaOH solution. The tests were performed at 303 and 353 K (30°C and 80°C). After an exposure duration of one month, the material consumption as well as the number and the depth of the pitting corrosion sites were determined.

Table 7 contains the test conditions and the results obtained at a test temperature of 303 K (30°C). The results for a test temperature of 353 K (80°C) are indicated in Table 8.

The results obtained at a test temperature of 303 K (30°C) show that, although the ratio of both components CO_2 or $(HCO_3)^-$ and their buffer capacity exert an influence on the pH value of the solution, a significant effect on the corrosion rates or pitting corrosion of aluminium cannot be found. At rising temperatures both the corrosion rates and the sensitivity to pitting corrosion clearly increase and also the influence of the pH value becomes more apparent.

Gas phase	$(HCO_3)^-$ ppm	pH	Corrosion rate $\mu m/a$	Number of pit holes per 40 cm^2	Depth of the pit holes μm
100% CO_2	0	4.5	3.4 ± 0.3	0	
	5	4.6	5.8 ± 0.5	0	
	300	5.4	0.5 ± 0.3	0	
Nitrogen with 1% CO_2	0	5.9	1.3 ± 0.3	0	
	5	6.1	0.8 ± 0.1	0	
	300	7.4	1.4 ± 0.2	0	
Nitrogen with 330 ppm CO_2	0	6.3	0.8 ± 0.1	2	80
	5	6.4	1.1 ± 0.1	0	
	300	8.5	3.6 ± 0.3	0	
Air	0	6.3	7 ± 1	5	160
	5	6.4	2 ± 1	1	140
	300	8.5	67 ± 4	3	120

Table 7: Results of the test of pure aluminium (Al99,5) exposed to 3 % NaCl solution at 303 K (30°C), test duration: 30 days [45]

Gas phase	(HCO$_3$)$^-$ ppm	pH	Corrosion rate µm/a	Number of pit holes per 40 cm^2	Depth of the pit holes µm
100% CO$_2$	0	5.2	41 ± 4	42	1070
	5	5.2	11 ± 4	23	775
	300	5.9	9 ± 1	1	720
Nitrogen with 1% CO$_2$	0	6.2	11 ± 1	3	120
	5	6.3	8 ± 1	3	50
	300	7.9	6 ± 1	0	
Nitrogen with 330 ppm CO$_2$	0	6.9	3 ± 1	1	360
	5	7.3	7 ± 1	1	825
	300	8.8	9.2 ± 0.5	0	
Air	0	7.0	18 ± 1	3	975
	5	6.7	6.2 ± 0.1	1	600
	300	8.9	10 ± 2	0	

Table 8: Results of the test of pure aluminium (Al99,5) exposed to 3 % NaCl solution at 353 K (80°C), test duration: 30 days [45]

A 3 Aluminium alloys

A maximum corrosion of 0.03 mm/a (1.18 mpy) was determined in corrosion tests over different periods (7, 28 and 70 days) performed in a rotating drum with the aluminium alloys AlMg, AlMn and AlMgSi in aqueous solution at 328 K (55°C) and at a partial carbon dioxide pressure of 13.8 bar [46].

Corrosion tests with the aluminium alloys EN AW-6063 (EN AW-Al Mg0.7Si) and EN AW-5052 (EN AW-Al Mg2.5) in carbonic condensates at 318 K (45°C) revealed corrosion rates of 0.07 mm/a (2.76 mpy) [47].

A 4 Gold and gold alloys

Only trace amounts of gold are dissolved in aqueous solutions of carbonic acid.

A 5 Cobalt alloys

The cobalt-chromium-tungsten alloys known as Stellite® contain 50 to 59 % Co, 26 to 33 % Cr, 4 to 13 % W and carbon contents above 1 %. They are recommended for the aboveground use in sour gas plants with carbon dioxide contents [48].

In these plants damage occurs, in particular, as a result of the interaction of hydrogen sulfide and moisture. Here, carbonic acid accelerates corrosion.

Since pure cobalt is hardly attacked by moist air and water or diluted hydrochloric acid and sulfuric acid [49], a major attack is not expected either in carbonic waters.

A 6 Chromium and chromium alloys

Hard chromium layers are very resistant to carbonic condensates [50]. Also chromium and chromium alloys exhibit a good resistance to carbonic acid.

A 7 Copper

Copper and copper base alloys are used in many applications due to their good corrosion behavior. Apart from their good corrosion resistance, their further application field in outdoor weather exposure, in fresh water pipelines, fittings, condensers, heat exchangers, in chemical plant construction and many other uses, is based also on their good processability, their good strength properties as well as their high thermal and electrical conductivity.

There is no data available about the behavior of copper towards carbon dioxide gas at higher temperatures, since this material is typically not used under such conditions. At ambient temperatures the use of copper pipes for transporting carbon dioxide, e.g. in medical technology, welding technology and in the beverage industry is permitted [51].

But copper is a material frequently used for pipes and other components exposed to water or aqueous solutions. However, in these cases carbon dioxide dissolved in the fluid may influence the corrosion behavior.

In contact with water the corrosion possibility of copper depends on the type of the cover layers formed by the corrosion products, which consist of Cu_2O or CuO depending on the type of fluid and the corrosion potential [52]. The more this layer prevents an ion and electron exchange between copper and water, the better is its protective effect and the higher is the resistance of copper. The carbon dioxide dissolved in water, the present alkaline earth ions and the carbonate/hydrogen carbonate system play an important role in the formation of the cover layer.

The influence of carbon dioxide in various waters on the corrosion of copper was investigated in connection with damage to drinking water pipes of copper [53]. Specimens from electrolytic copper and hard-drawn copper pipes and the following test solutions were used for the tests:

1. Distilled water with 5 mg/l CO_2 and 10 mg/l SO_4^{2-} added by dissolving the adequate amounts of sodium hydrogen carbonate and calcium sulphate. In several tests 5 mg/l iron in the form of iron(III) chloride were also added to the solution.
2. Drinking water from the Oslo waterworks and water from the supply system purified by means of an ultra-membrane filter. The membrane filter reduced

the content of organic carbon compounds to 2.5 mg/l C and the iron content to < 0.02 mg/l Fe.
3. A test solution free from organic substances and manganese, produced by dissolving inorganic salts in distilled water with the same concentrations as in test solution 1.

The tests were performed in a circulation apparatus under thermostatically controlled conditions with a flow rate of 50 ml/h. One, two or three small inflow tubes were provided to pass different, automatically controlled amounts of CO_2 gas into the solution. The contents of free and bound carbon dioxide were calculated using equilibrium equations. Figure 9 depicts the variation of the carbon dioxide content over time and the mean values.

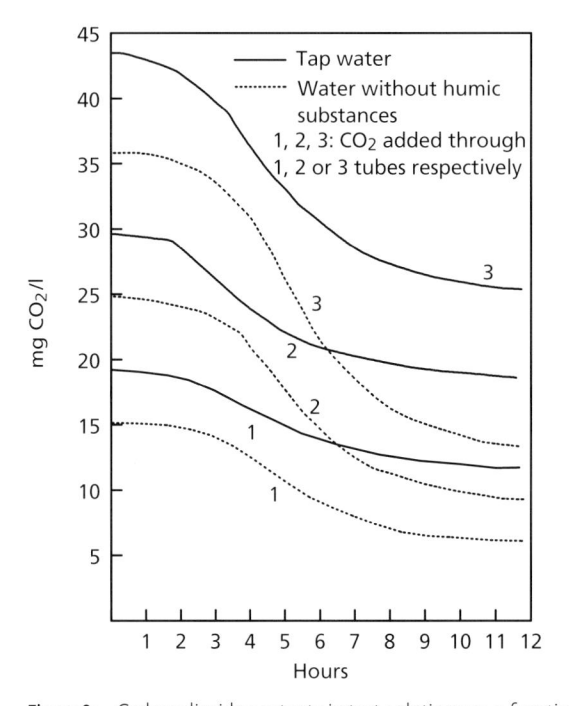

Figure 9: Carbon dioxide contents in test solutions as a function of time and mean contents [53] Mean values of the CO_2 content (mg/l) for tap water: 1 = 14,6; 2 = 22,8; 3 = 32,9. For water without organic substances: 1 = 10,0; 2 = 16,3; 3 = 23,9.

The tests were carried out at temperatures of 303 to 343 K (30°C to 70°C). Figure 10 shows the corrosion rates obtained after a test period of three months from tests in the test solution 2 with and without organic substances as a function of the test temperature. In both solutions the addition of carbon dioxide lead to a clear increase in the corrosion rates. Also the presence of organic substances exerted a negative influence on the corrosion of the copper specimens. The lower three curves of the figure might be indicative of a critical temperature in the range of 323 K (50°C).

Figure 11 depicts the test results in both test solutions after the addition of different amounts of carbon dioxide through one, two or three of the small inflow tubes. A rising carbon dioxide content leads to a clear increase in the corrosion rates of the copper specimens in both solutions.

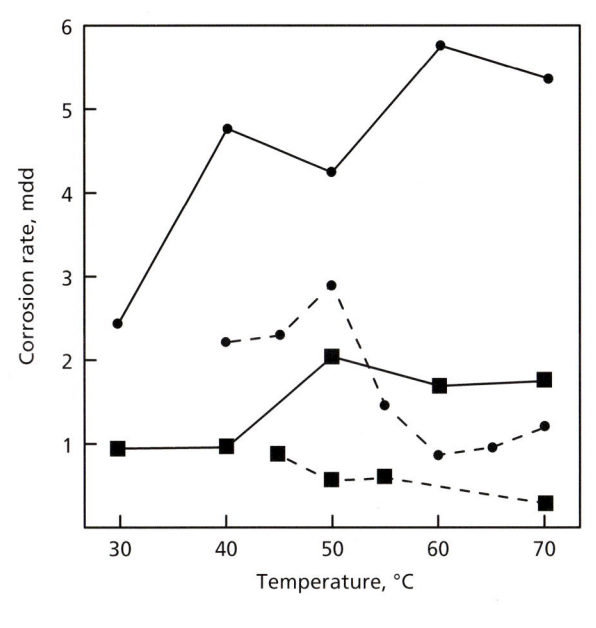

(mdd = mg/dm² d)   1 mdd x 0.004 = 1 g/m² h)

Figure 10: Corrosion rates of copper after three months as determined in test solution 2 with and without organic substances as a function of the test temperature [53]

The influence of dissolved carbon dioxide in drinking water on the corrosion of pure copper was also confirmed by investigations in Sweden [54]. These investigations included exposure tests as well as potentiodynamic measurements with rotating disk electrodes in synthetic drinking water with different contents of hydrogen carbonate, calcium and chloride. Double distilled water with $NaHCO_3$ additions of concentrations between 0.5 and 10 mM was chosen as the synthetic drinking water. The pH value was adjusted either by passing CO_2 into the solution or by adding HCl. Calcium was added as calcium nitrate and chloride as NaCl.

In addition, drinking water samples from copper pipes were randomly taken at various sites and various times and analyzed to determine their content of dissolved copper, calcium and carbon dioxide. The samples were taken each once from water having stagnated over night and once from water after having kept running for 10 minutes (20 l/min). Figure 12 and Figure 13 show the share of water samples with a content of dissolved copper of more than 300 µg/l for the content of free CO_2 and for the ratio of free CO_2 and calcium ions, respectively.

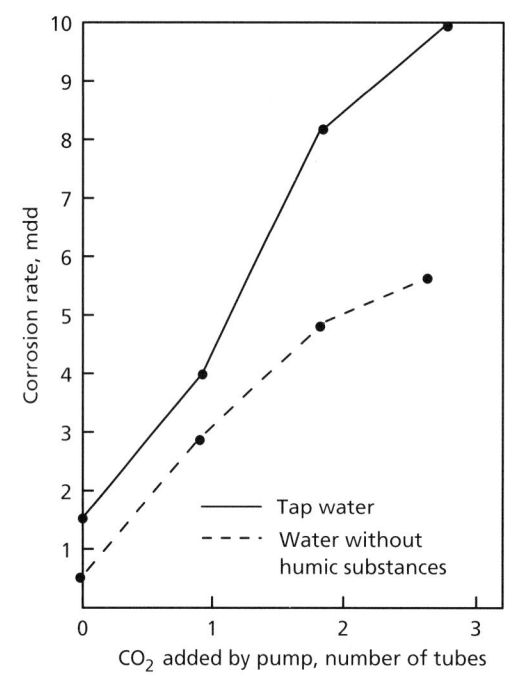

Figure 11: Corrosions rates of copper in drinking water with different carbon dioxide contents [53]

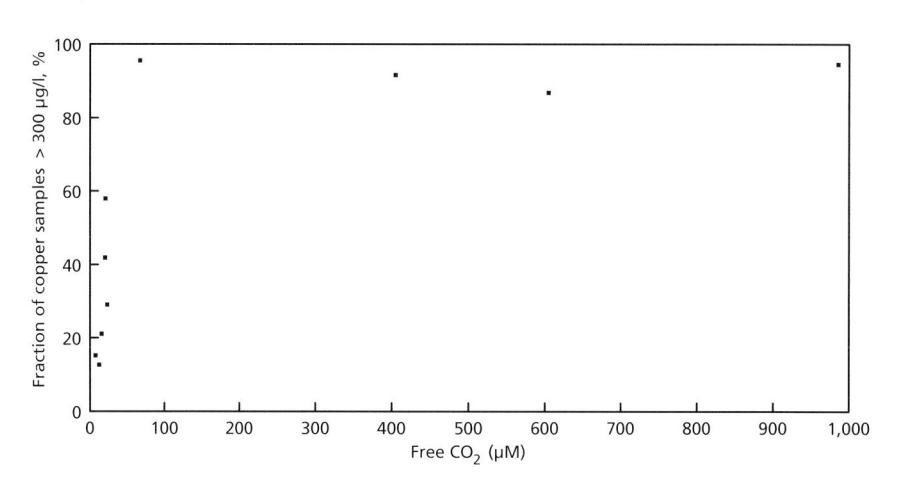

Figure 12: Percentage of water samples with copper contents higher than 300 µg/l as a function of the content of free CO_2 [54]

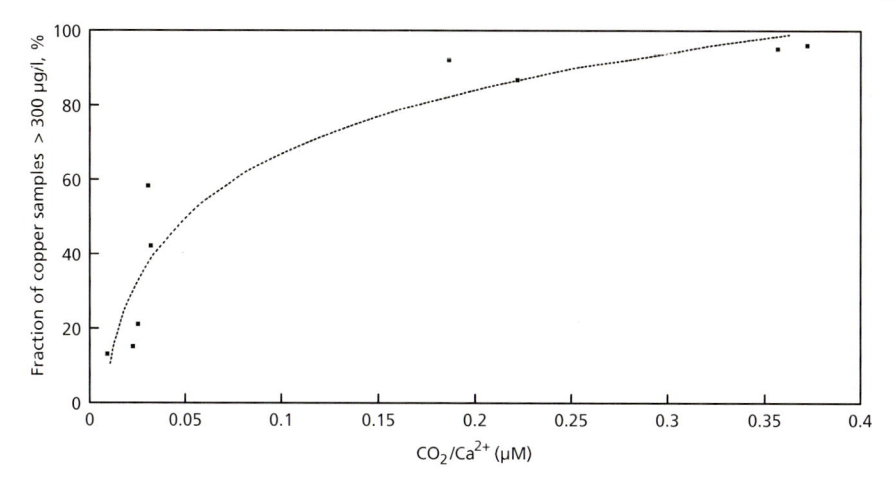

Figure 13: Percentage of water samples with copper contents higher than 300 µg/l as a function of the ratio of free CO_2 and calcium ions [54]

Comparison of both evaluations reveals that the decisive parameter for copper corrosion in drinking water is the ratio of free CO_2 and calcium ions rather than the content of free carbon dioxide. Figure 14 depicts the results of the electrochemical investigations in synthetic seawater and confirms the effect shown in Figure 13.

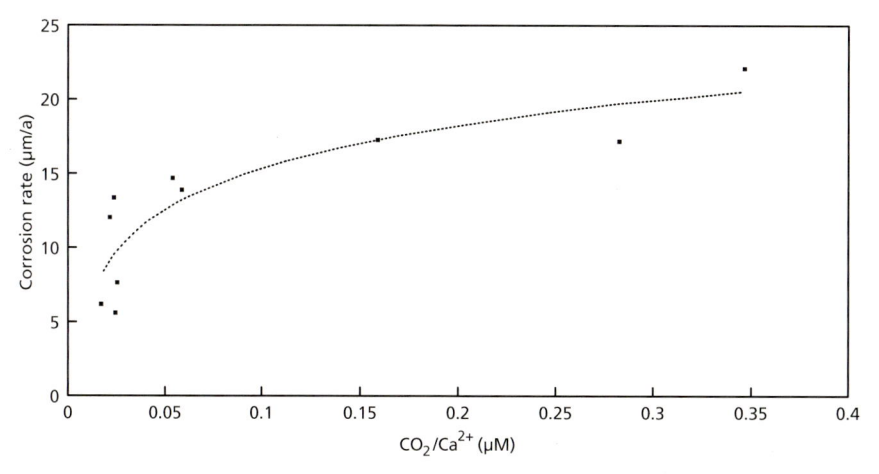

Figure 14: Dependence of the corrosion rate of pure copper in synthetic drinking water on the ratio of free CO_2 and calcium ions [54]

Considering the carbonate system in the absence of calcium, the content of free CO_2 is determined by the pH value and the alkalinity. In Figure 15 the copper content determined in solution following the exposure of the copper specimens to synthetic drinking water as a measure of corrosion is plotted logarithmically against the CO_2 content.

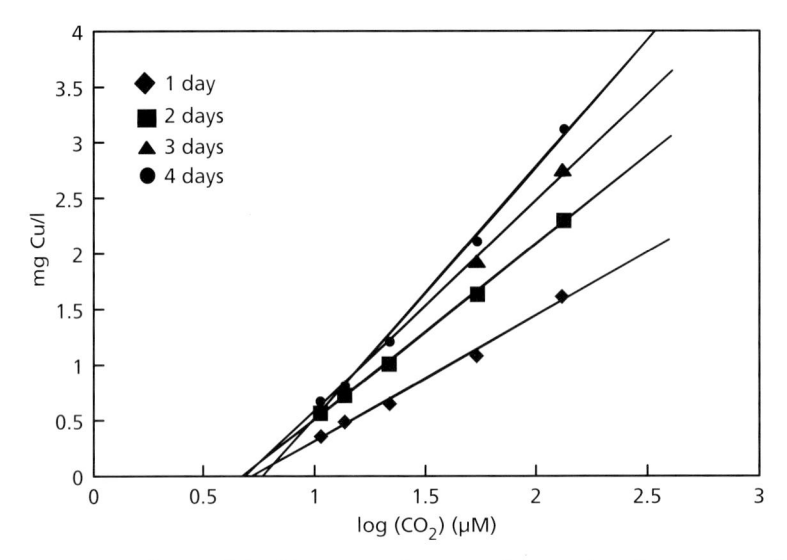

Figure 15: Influence of the CO_2 content of water on the corrosion of pure copper after various exposure periods [54]

Also the European Standard DIN EN 12502-2 provides information about investigation results and experience regarding the influence of free carbon dioxide and the alkaline earth content in waters obtained from the corrosion of copper materials in drinking water distribution systems of buildings [55]. By analogy this standard can be applied to other water systems. Under normal conditions drinking water systems of copper and copper alloys are generally resistant to corrosion damage. However, there are certain conditions under which corrosion damage may occur.

The most frequent manifestation is pitting corrosion of copper in water pipes. A distinction is made between two forms of pitting corrosion of copper in **water pipe systems**. The type of pitting corrosion depends on the water temperature, the composition of the water and the operating conditions.

Pitting corrosion – type 1 occurs almost exclusively in cold water pipes and is mainly found in moderately hard ground waters and less often in surface waters, which may contain organic substances as potential inhibitors [56].

Typically, the site of attack is characterized by semi-spherical pits with overlying green pustules mainly consisting of basic copper carbonate. Beneath these pustules the pits are always covered by a coherent copper(I) oxide layer. Beneath this layer there are copper oxides and chlorides. Regarding the effect of the water composition, a major role is plaid by the ratio of alkaline earth salts (chlorides, sulfates, nitrates) and alkaline earth hydrogen carbonates. The probability of corrosion decreases as the concentration of hydrogen carbonate ions increases and the concentration of alkaline earth ions decreases. These relations may be described by using a factor Q, whereas:

$$Q = \frac{2 \times c \text{ (alkaline earth ions)}}{c \text{ (hydrogen carbonate ions)}} - 1 \, \text{Mol/m}^3$$

There is a low risk of pitting corrosion if Q < 0,15 and Q > 2.
There is a higher risk of pitting corrosion if Q = 0,5 to 1.

Comprehensive investigations into the influence of the water composition on the occurrence of pitting corrosion in water pipes of copper covering a period of almost 10 years confirm the influence of the ratio of alkaline earth salts and alkaline earth hydrogen carbonates described in said formula [57]. In addition, the corrosion values calculated from this relation and applicable to the evaluation of water depend on the temperature, the content of free carbonic acid and the alkali content.

Pitting corrosion – type 2, which occurs less frequently than type 1, is characterized in that the corroding pipes appear to be intact. Often the inside is covered by pale yellow to yellow-brown amorphous deposits [56]. Beneath there are spatially very limited sites of pitting corrosive attack with irregular internal geometries, leading to holes in the form of pinpricks to the outer side of the tubes if they break up. The pitting corrosion sites are completely filled with copper(I) oxide.

Type 2 pitting corrosion mainly occurs in warm water, in particular in soft and acidic waters (pH 6 to 4.2). Often, those waters encouraging type 2 pitting corrosion contain higher contents of aggressive carbonic acid and, seldom, the carbonate hardness is higher than 1° dH. Also the sulfate content exerts an influence, with the molar ratio of hydrogen carbonate/sulfate playing a role here.

At a molar ratio:

$$S = \frac{c \text{ (hydrogen carbonate ions)}}{c \text{ (sulfate ions)}} \geq 2$$

or at pH values > 7.5 the probability of pitting corrosion is low [55].

As suggested by the investigations described in [58], the contents of aggressive carbonic acid are below 10 mg/l if type 1 pitting corrosion occurs, whereas these contents amount up to 50 mg/l for type 2.

Contrary to the general knowledge that hydrogen carbonates in waters are favorable to counteract the corrosion of copper, [59] suggests that hydrogen carbonate may also have a detrimental effect under certain conditions. Results obtained from measurements of test tubes in the lab and from investigations performed in drinking water supply installations were used to evaluate the connection with water conditions and the release of copper corrosion products. Results obtained from measurements of test tubes in the lab and from investigations performed in drinking water supply installations were used to evaluate the connection with water conditions and the release of copper corrosion products. As turned out, the corrosion rates of copper and the release of corrosion products with higher contents of hydrogen carbonates are higher in fairly new copper pipes. At a constant pH value there is a linear increase of the concentrations of copper corrosion products with increasing bicarbonate contents. It is assumed that this phenomenon is attributable to the increase of solubility of copper hydroxide caused by the bicarbonate ions. This effect diminishes at higher pH values. Therefore, desorption of the dissolved carbon dioxide by aeration and hence a higher pH value are recommended as a countermeasure.

Also in steam systems dissolved carbon dioxide or carbon dioxide carried along in occurring condensates may cause a considerable corrosive attack on copper components and therefore care must be exercised as to carefully treat the water [60].

A 8   Copper-aluminium alloys

Older publications in the literature indicate material removal values of 0.006 to 0.02 mm/a (0.24 to 0.79 mpy) for copper-aluminium alloys in carbon dioxide-containing waters depending on the composition of the alloy. In practice these materials are well established also in humid flue gases with carbon dioxide as well as in carbonic mineral waters.

A 9   Copper-nickel alloys

As already mentioned in A 7, dissolved carbon dioxide in steam systems or in condensates may cause damage to copper components. [61] describes an incidence of damage to pipes of a copper-nickel material CuNi90/10 (corresponding to CW352H, CuNi10Fe1Mn) in the condenser of a thermal seawater desalination plant. Non-condensable gases, such as CO_2, O_2, N_2, NH_3, from the steam and air entering the system through leaks accumulate in the condensers of such plants. Here, carbon dioxide may originate from both the decompensation of hydrogen carbonates and the transport of CO_2 from the salt solution into the steam. In such gas-loaded condensates the protective effect of the oxide layers formed on the copper materials is not sufficient to prevent a corrosive attack. In the present case the damage was attributed to the dissolved carbon dioxide in the condensate together with low amounts of sulfur-containing compounds. As a corrective measure it is recommended to provide quick and effective ventilation in order to expel the dissolved gases. During periods of downtime the tubes should be flushed and dried to remove the remaining gas-saturated condensate remnants and facilitate the formation of protective copper oxide layers.

A 10   Copper-tin alloys (bronze)

For copper-tin alloys in distilled water saturated with carbon dioxide in a temperature range from 303 to 323 K (30°C to 50°C) the indicated material removal values amount from 0.007 to 0.02 mm/a (0.28 to 0.79 mpy).

The information sheets of the German Copper Institute refer to a good corrosion resistance of various copper-tin alloys against carbonic acid. According to these sheets also the good behavior in the atmosphere is not affected by the contents of sulfur dioxide and carbon dioxide [62].

A 14 Unalloyed and low-alloy steels/cast steel
A 15 Unalloyed cast iron and low-alloy cast iron

Behavior in gaseous carbon dioxide

The behavior of unalloyed and low-alloy steels as well as of high-temperature and heat-resistant steels in hot carbon dioxide gas is of interest in connection with their use in exhaust gases and in nuclear reactors cooled with CO_2 gas, and hence has been thoroughly investigated.

Carbon dioxide is the only medium known to be able to cause local scale-offs and scale nodules in unalloyed and low-alloy steels under pressure and at high temperatures. These scale-offs may occur even after a long time. This needs to be observed when performing investigations and evaluating results. When plotting the mass increase against time for such tests, a curve is typically obtained with three different sections as schematically shown in Figure 16.

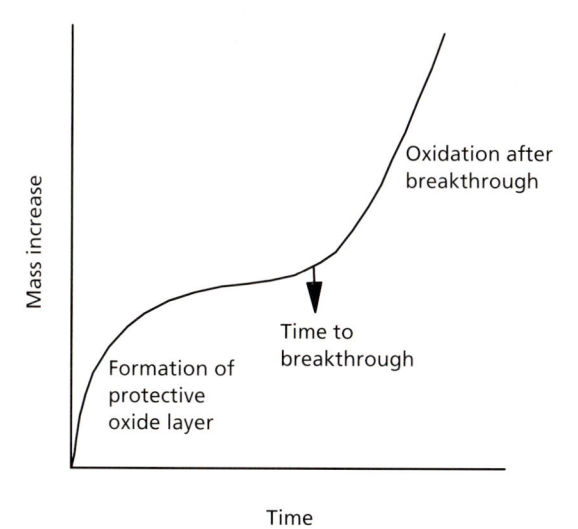

Figure 16: Schematic curve of mass change during the oxidation of low-alloy steels in carbon dioxide at high temperatures and pressures [63]

In section 1 a protective layer of magnetite Fe_3O_4 is formed and the oxidation process follows an almost parabolic time law. In the transitional section 2 local break-offs occur in this magnetite layer, increasing in number and size in the course of time until the protective layer on the entire surface is finally destroyed. Thereafter, the oxidation process follows a linear time law in the third section.

This oxidation process has caused substantial damage to heat exchanger tubes and cladding tubes for fuel assemblies in nuclear reactors cooled with CO_2 gas.

The time until a breakthrough occurs decreases as

- the temperature increases,
- the pressure increases,
- the water vapor content of the carbon dioxide increases,
- the surface roughness increases.

In unalloyed and low-alloy steels the water vapor content in carbon dioxide exerts a clear effect on the oxidation behavior. It does not only shorten the time until a breakthrough occurs and hence supports the formation of non-protecting cover layers, but also influences the oxidation rate after a breakthrough. Lowering the water vapor content from 25 ppm to 1 ppm doubles the time till the breakthrough and at water vapor contents of less than 10 ppm scaling follows a parabolic time law. At higher water vapor contents the time dependence of the oxidation rate is almost a straight line with a gradient increasing as the water vapor content increases as shown in Figure 17 using the example of the temperature resisting steel 13CrMo4-5 (DIN Mat. No. 1.7335).

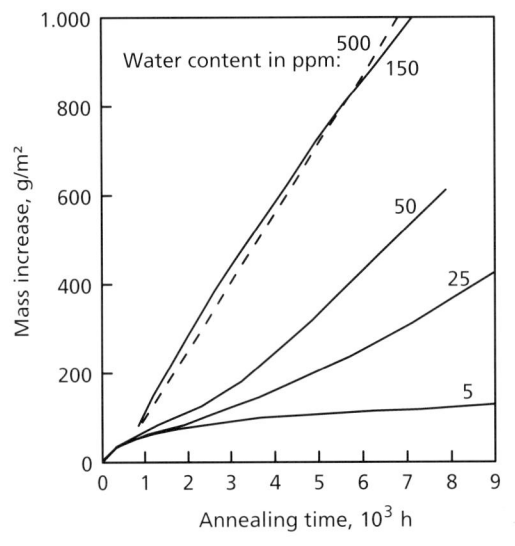

Figure 17: Oxidation behavior of the temperature resisting steel 13CrMo4-5 as a function of time and the water vapor content of carbon dioxide [64]

The formed scale always contains a considerable amount of carbon, the carbon content being especially high at the steel/scale phase boundary. The carbon content increases as the oxide layer thickness and the water vapor content in the carbon dioxide increase. The carbon content in the broken off layers (section 2 in Figure 16) is higher than in the initially formed protecting layers (section 1 in Figure 16).

Regarding the mechanism for breaking off the oxide layer it is assumed that the initially formed magnetite layer grows due to the migration of iron ions and electrons from the inside to the outside. This leads to the formation of voids at the steel/

Fe$_3$O$_4$ interface, penetrated by CO$_2$ along microcracks. Due to the low oxygen pressure in the pores a CO-CO$_2$ mixture is formed, which in redox reactions transports oxygen from the compact oxide layer to the steel surface so that the oxidation process will continue.

This mechanism also serves as an explanation for the formation of scale with two layers, which could always been found. Beneath an external compact Fe$_3$O$_4$ layer there is a porous Fe$_3$O$_4$ layer in contact with the steel surface, where alloying elements and minor ingredients of the steel accumulate. However, in equilibrium with Fe/Fe$_3$O$_4$ a CO-CO$_2$ mixture is thermodynamically instable below 873 K (600°C) and can deposit carbon according to Boudouard's equilibrium (Equation 9).

During scale tests with intermediate cooling higher mass increases compared to isothermal tests are only obtained in the beginning. However, after a longer test duration significant differences cannot be found since it is only the upper scale layers spalling during a temperature change, whereas the metal-near layers determining the rate are not removed.

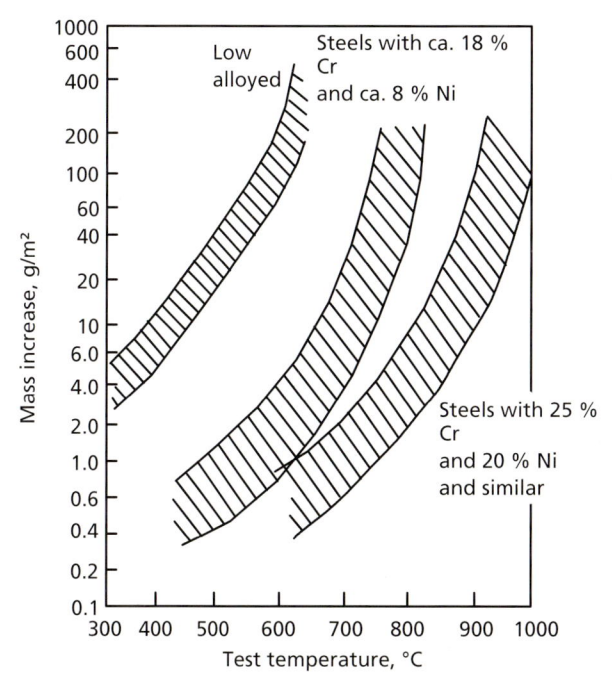

Figure 18: Mass increase of various steel groups in isothermal and cyclic scaling tests in pure dry carbon dioxide (test duration 1000 h) [64]

Figure 18 summarizes the results of isothermal and cyclic scaling tests for several steel groups in pure dry carbon dioxide [64].

The scatter band of each steel group is rather narrow. The low-alloy, high temperature ferritic steels containing molybdenum, chromium or molybdenum and chromium can be used up to temperatures of about 723 K (450°C). Unalloyed steels fall under the scatter band of these low-alloy steels up to a temperature of about 673 K (400°C), however their scaling rate is higher at higher temperatures and they exhibit breakthroughs of the scale layer at an early stage.

The influence of the flow rate of the gas on oxidation depends on the temperature. At flow rates up to 60 m/s below 723 K (450°C) an effect could not be established for the low-alloy chromium-molybdenum steels. From 763 K (490°C) oxidation increases as the flow rate increases, in particular due to the fact that the spalling intensifies at rates from about 50 m/s.

For smelting steel in the converter with a higher input of scrap metal it is important to avoid oxidation of the iron with carbon dioxide, carbon monoxide, hydrogen and water vapor, which occur when fossil fuels are burnt. Studies in this connection involved investigations into the kinetics of the iron oxidation in carbon dioxide and in carbon dioxide/carbon monoxide mixtures in a temperature range from 1573 to 1723 K (1300°C to 1450°C) [65].

The oxidation tests were carried out in a Tamman furnace and the mass changes of the specimens were monitored by means of a thermogravity balance. The reaction gases were preheated to 1373 K (1100°C) in a preheating furnace. Since the formed iron oxide smelts at test temperatures above 1650 K (1377°C), the oxide dripping down was caught in a ceramic vessel. Figure 19 depicts the test apparatus used.

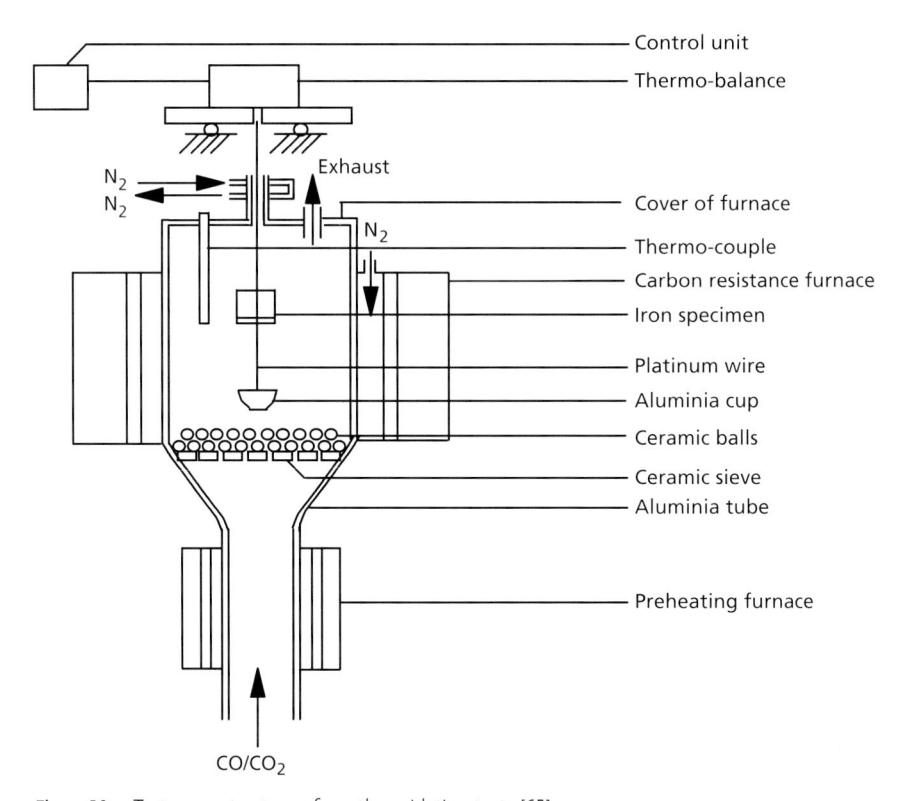

Figure 19: Test apparatus to perform the oxidation tests [65]

Figure 20 shows typical curves describing the time dependence of the mass increase of specimens for various carbon dioxide/carbon monoxide mixtures at 1597 K (1324°C) and a constant flow rate of 20 cm/s of the gases.

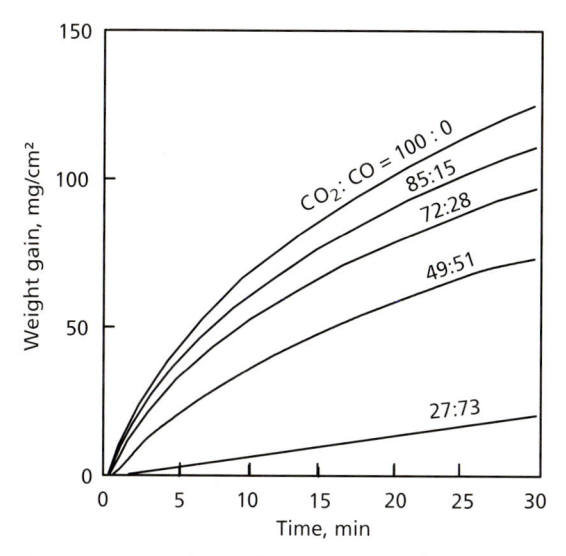

Figure 20: Dependence of mass increase of iron specimens in CO_2-CO mixtures at 1597 K (1324°C). Flow rate 20 cm/s [65]

If the carbon dioxide content is high, the curves initially follow a linear and subsequently a parabolic time law. If the carbon dioxide content is lower, the range of the linear time dependence covers a clearly longer period compared to high carbon dioxide contents. The mass increase is clearly reduced as the carbon dioxide concentration decreases.

In Figure 21 the oxidation curves of pure carbon dioxide are plotted against a constant temperature of 1673 K (1400°C) and various flow rates.

The time dependence of oxidation is a linear curve since at this temperature the formed iron oxides smelt and drop off and inhibition of the oxidation processes is not possible.

Figure 22 shows the oxidation curves obtained in pure carbon dioxide at temperatures between 1373 and 1673 K (1100°C and 1400°C)

At temperatures up to 1633 K (1360°C) a parabolic time dependence is observed after a short time and the reaction product mainly consists of wustite. At higher temperatures liquid iron oxides are formed and, therefore, the oxidation curve follows a linear time law as can be seen already in Figure 21.

Comprehensive investigations into the reaction mechanism and the kinetics of the oxidation of iron in carbon dioxide/carbon monoxide mixtures are also described in [66, 67]. Here, specimens with a thickness of 0.25 to 2 mm consisting of high-purity iron (99.99%) were examined in CO/CO_2 gas mixtures of various compositions at temperatures of 1273 to 1473 K (1000°C to 1200°C) at total pressures of 0.1

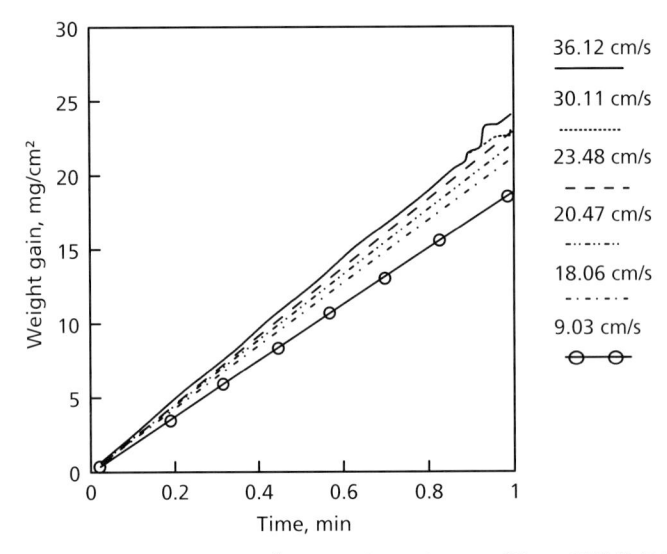

Figure 21: Mass increase of iron specimens in pure CO_2 at 1673 K (1400°C) and various flow rates [65]

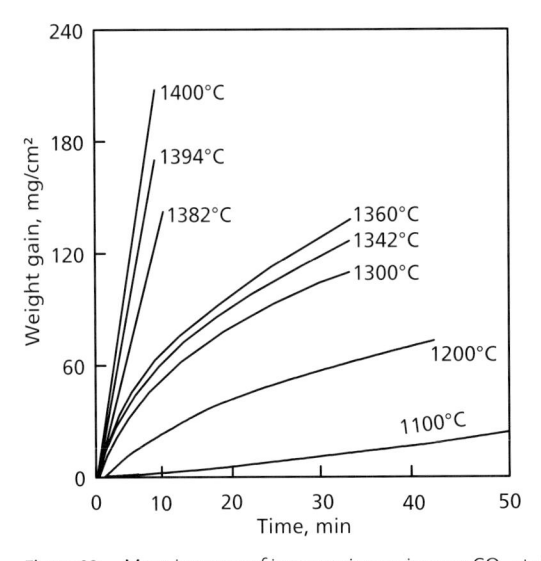

Figure 22: Mass increase of iron specimens in pure CO_2 at different temperatures [65]

to 1.0 bar. The oxidation rates were thermogravimetrically determined. Figure 23 shows the influence of the flow rate of the test gases. At a test temperature of 1473 K (1200°C) there is practically no dependence of the oxidation rate on the flow rate if the flow rate is equal or higher than 1 cm/s. At a lower flow rate the oxidation rate decreases, which is attributable to insufficient access of the reaction gases to the surface. At the lower test temperature of 1273 K (1000°C) smaller flow rates are sufficient to reach a constant oxidation rate.

Figure 24 shows the connection between the mass increase of the iron specimens and the test temperature, the composition of the test gases and the test duration for the onset of the oxide formation and the first growth of the oxide layer.

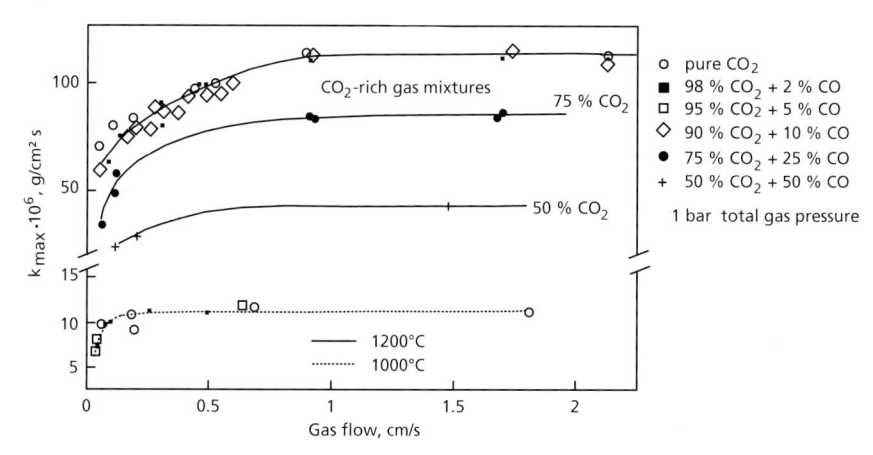

Figure 23: Maximum oxidation rate (k) of pure iron in different CO/CO_2 gas mixtures at a total pressure of 1 bar and temperatures of 1273 and 1473 K (1000°C and 1200°C) as a function of the gas flow rate [66]

The upper part of the figure shows that at a low total pressure of the gas the section of the linear rise of the mass increase over time is preceded by a period during which the reaction rate is lower. This induction phase attributable to the nucleation of oxides is shorter the higher the temperature and the CO_2 content in the gas are. At 1473 K (1200°C) the induction phase is so short that the regions of film growth and subsequent increasing oxidation rate overlap. This also holds true for the oxidation process at a higher total pressure as the lower part of the figure shows for a pressure of 1 bar.

If the test duration is longer, a linear rise of the mass increase of the specimens over time can be observed in practically all cases.

As can be seen in Figure 26, the maximum oxidation rate increases as the total pressure of the gas and the content of carbon dioxide in the test gas increase.

Behavior in aqueous carbon dioxide solutions

The behavior of the unalloyed and low-alloy steels in aqueous carbon dioxide solutions is of special technical importance in connection with the extraction, storage, transport and processing of crude oil and natural gas since the products to handle there almost always exhibit an aqueous CO_2-containing phase. CO_2 is already contained in the extracted products, in particular from deeper reservoirs, or it is

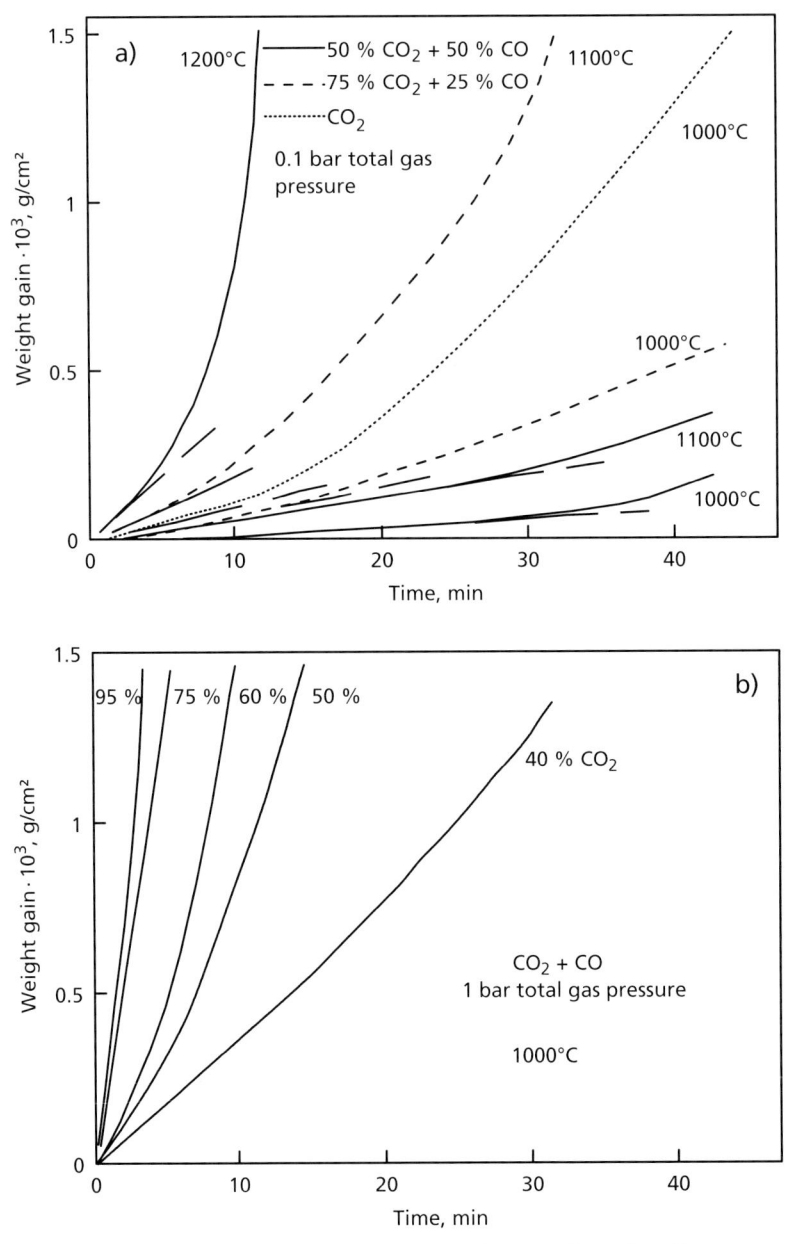

Figure 24: Mass increase of the iron specimens as a function of time for different compositions, temperatures and pressures [66]

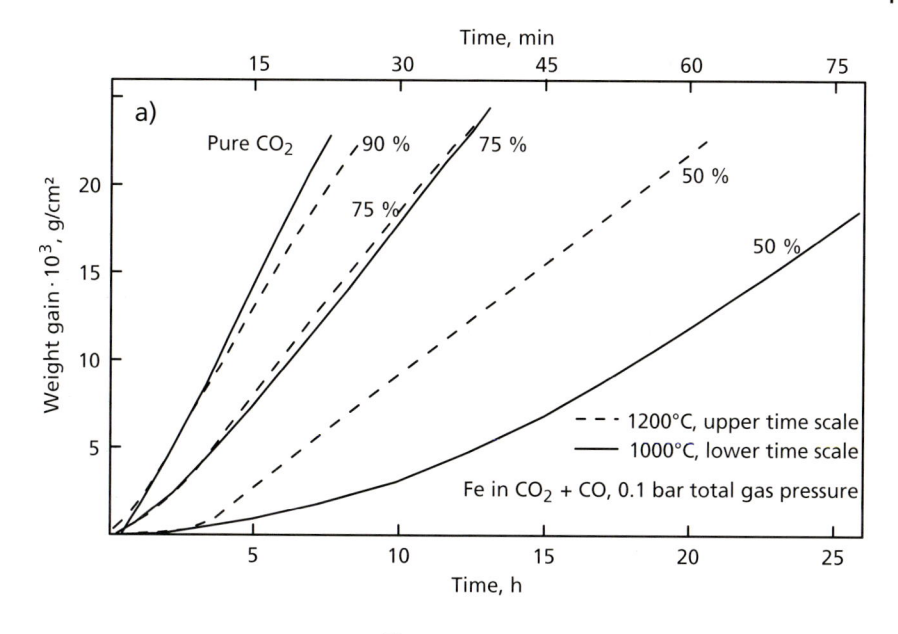

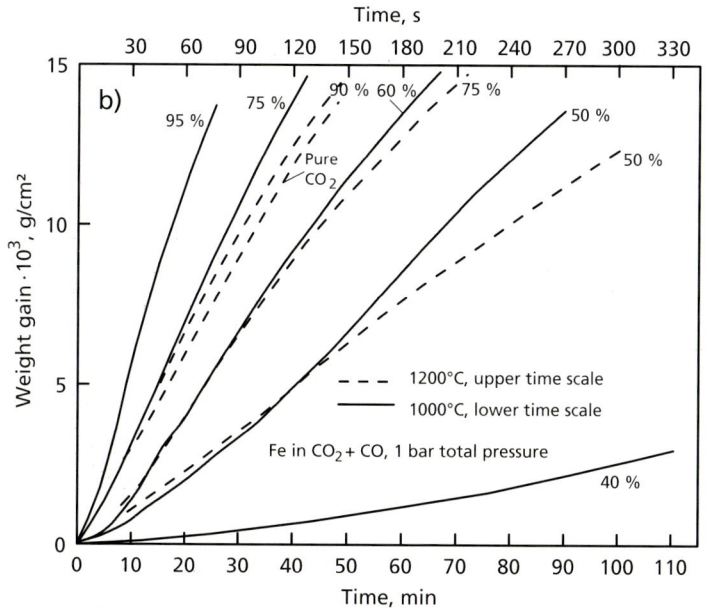

Figure 25: Mass increase of the iron specimens as a function of time for different compositions, temperatures and pressures as well as a longer test duration [66]

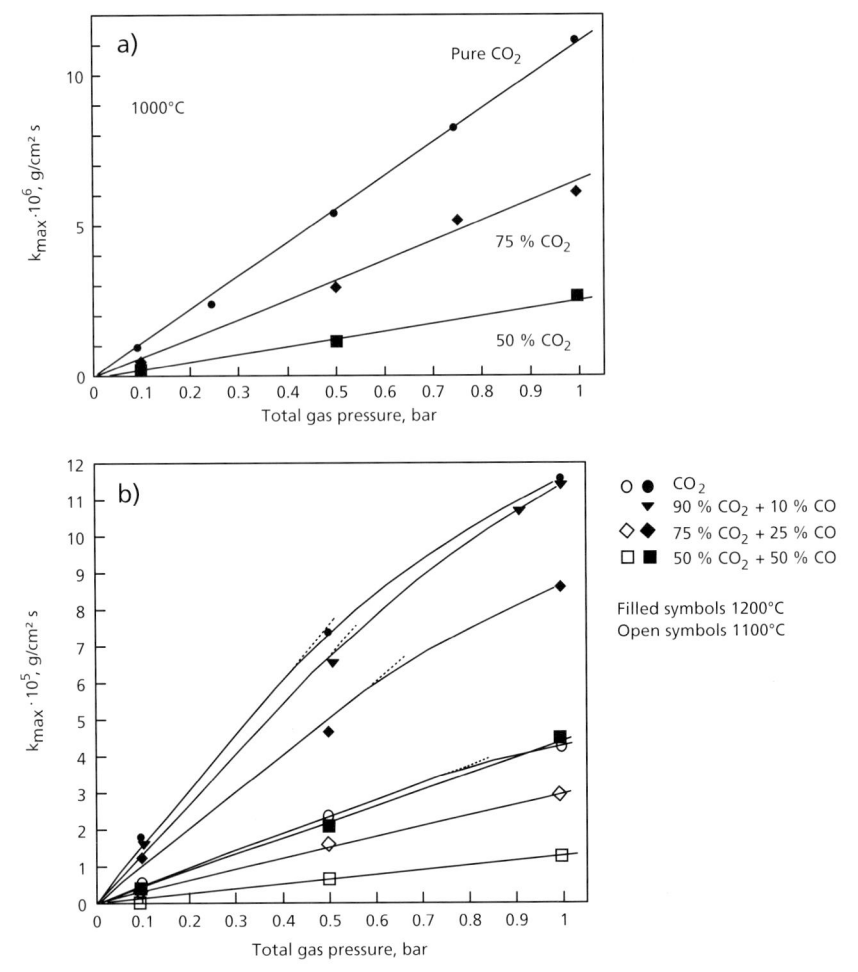

Figure 26: Maximum oxidation rates (k) of the pure iron specimens as a function of the total gas pressure at different temperatures and gas compositions [66]

injected into the reservoirs if the pressure drops. Correspondingly, there are many publications in the literature dealing with the mechanism of steel corrosion in aqueous carbon dioxide solutions [68–74].

CO_2 dissolved in aqueous fluids reacts according to Equation 1, forming the weak acid H_2CO_3. As can be seen from Figure 4 and Figure 5, the dissolved carbon dioxide causes the pH value of the solutions to decrease, and therefore the corrosion of steel in CO_2-containing oxygen-free waters occurs as an acid or water type corrosion. The entire reaction can be described with

Equation 12 $\qquad Fe + 2\,CO_2 + 2\,H_2O \rightarrow Fe^{2+} + 2H_2 + 2\,HCO_3^-$

The rather strong corrosive attack of CO_2 containing oxygen-free waters on steel is not solely attributed to the low pH value. It is known that the attack of carbon dioxide solutions on steels is stronger than that in diluted mineral acids with the same pH value [68, 69]. This is attributed to the hydrogen evolution intensified by carbon dioxide in the cathodic part process of the corrosion reaction. Hydrogen can be formed by the reduction of non-dissociated carbonic acid adsorbed on the metal surface by reduction [69]:

Equation 13 $H_2CO_3 \text{ (ad)} + e^- \rightarrow H \text{ (ad)} + HCO_3^-$

For details regarding the various mechanisms and individual steps reference is made to the cited literature.

The corrosion products formed during the reaction of the steel with the carbonic acid in oxygen-free solutions mainly consist of iron carbonate and offer a certain protection against further corrosion. The composition and structure of the layers depend on the steel grade and the surrounding conditions [75–78].

The corrosion of steel in oxygen-free CO_2-containing solutions is characterized in that the cathodic hydrogen evolution does not only occur in the form of a reduction of hydrogen ions, but also by direct reduction of adsorbed carbonic acid formed by hydration of CO_2. Here, the corrosiveness is a function of the partial pressure of CO_2 and the temperature. Basically, there is a risk of erosive corrosion above 0.5 bar. Temperature increases exert an especially noticeable influence. Freely corroding steel in CO_2-containing solutions absorbs almost double the amount of corrosion hydrogen compared to that in a sulfuric acid at identical pH values [79].

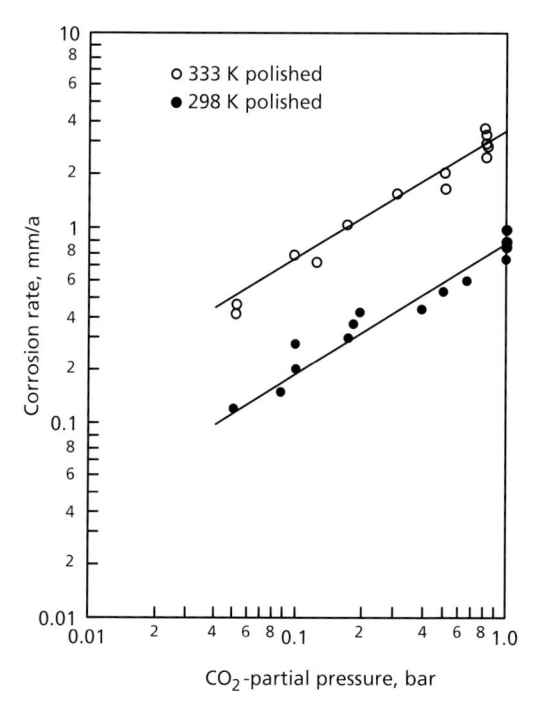

Figure 27: Influence of the partial pressure of CO_2 on the corrosion rates of unalloyed steel in 0.1 % NaCl solution at 298 and 333 K (25°C and 60°C) [70]

The investigations described in [70] determined the influence of the partial pressure of carbon dioxide on the corrosion of polished specimens of unalloyed steel in a 0.1% NaCl solution at 298 and 333 K (25°C and 60°C) as shown in Figure 27. More results for different temperatures and partial pressures of carbon dioxide are compiled in Table 9.

Temperature K (°C)	Partial pressure of CO_2 bar	Corrosion rate mm/a (mpy)
278.5 (5.5)	1	0.4 (15.7)
285 (12)	1	0.6 (23.6)
288 (15)	1	0.7 (27.6)
	0.52	0.4 (15.7)
	0.37	0.3 (11.8)
	0.21	0.2 (7.87)
295 (22)	2	1.6 (63.0)
	1	0.9 (35.4)
303 (30)	1	1.3 (51.2)
313 (40)	0.92	1.7 (66.9)
323 (50)	0.88	2.3 (90.6)
333 (60)	0.80	3.9 (154)
343 (70)	0.69	4.3 (169)
353 (80)	0.53	5.7 (224)

Table 9: Influence of the partial pressure of CO_2 and the temperature on the corrosion rate of unalloyed steel in 0.1 % NaCl solution [70]

The high dependence of the corrosion rate of unalloyed steel in carbon dioxide-containing solutions on the content of dissolved CO_2 was also confirmed by electrochemical measurements in sodium sulfate solutions [71].

Figure 28 shows the corrosion current densities measured at rotating disk electrodes from the unalloyed steel, DIN Mat. No. 1.1632, in 0.5 M sodium sulfate solution as a function of the dissolved carbon dioxide at 298 K (25°C).

During the investigations described in [72] the influences were analyzed of lower alloy contents as well as of the purity degree of the carbon dioxide flushing gas on the corrosion of unalloyed steels in 0.5 M sodium sulfate solution at 298 K (25°C). The chemical analytical values of the examined steels are indicated in Table 10.

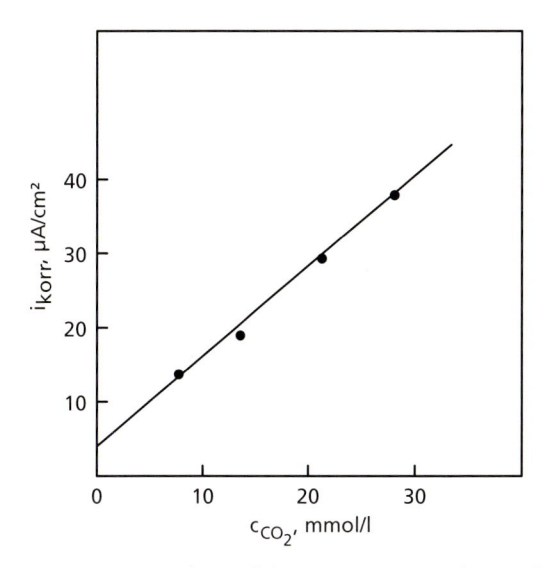

Figure 28: Dependence of the corrosion current density of unalloyed steel on the CO_2 content in an 0.5 M Na_2SO_4 solution at 298 K (25°C) [71]

No.	Mat. No.	C	Si	Mn	P	S	Al	Cu	Cr	Ni	N
1	1.1623	0.067		0.380	0.016	0.016			0.028	0.024	0.0015
2	1.0375	0.040	0.020	0.210	0.016	0.020	0.035	0.020	0.010	0.030	0.0074
3	1.1003	0.020	0.010	0.010	0.002	0.009	<0.001	<0.01	0.010		0.0030
4	1.1104	0.004	0.010	0.190	0.010	0.023		0.030		0.080	
5	1.8961	0.090	0.290	0.580	0.016	0.025	0.057	0.520	0.640	0.270	
6	1.8962	0.090	0.430	0.270	0.080	0.023	0.037	0.400	0.560		0.0090
7	x)	0.011		<0.01	0.006	0.005		0.006	0.145		0.012
8	x)	0.010		<0.01	0.059	0.005		0.003	<0.002		0.009
9	x)	0.008		<0.01	0.010	0.005		0.175	<0.002		0.010
10	x)	0.009		<0.01	0.004	0.005		0.163	<0.002		0.010

x) Special smelts

Table 10: Chemical analytical values of the examined steels [72]

The test solution was flushed with carbon dioxide until the desired CO_2 content was reached in the solution, while the maximum test duration was 500 hours. The material consumption of the rotating disk specimens was determined at intervals of 25, 50, 100, 250 and 500 h. To examine the influence of a low oxygen content in the carbon dioxide gas, apart from the standard CO_2 gas with 100 ppm O_2 also ultrapure carbon dioxide was used with an oxygen content of < 0.1 ppm O_2. The material removal values determined for steel 1 as a function of the test period in 0.5 M sodium sulfate solution with different standard CO_2 contents are plotted in Figure 29. The material consumption rates increase as the test period increases and, in particular, as the carbon dioxide content in the solution increases. But strikingly, lower material consumptions were measured for the highest carbon dioxide content of 27 mmol/l in the solution compared to the lower value of 21 mmol/l. This reproducible result is attributed to the fact that increasingly protective cover layers are formed at high carbon dioxide contents.

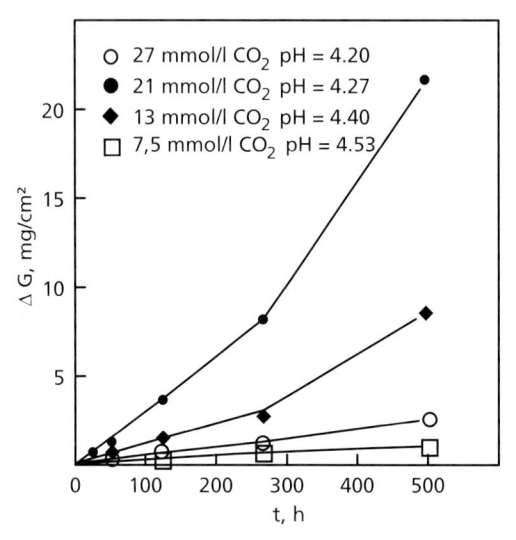

Figure 29: Time dependence of the material consumption rates of steel 1 in 0.5 M sodium sulfate solution with different standard CO_2 contents [72]

The corresponding measurement results for the tests with oxygen-free carbon dioxide are indicated in Figure 30.

The material removal values are clearly lower than those of the oxygen-containing solution; but here again lower material consumptions are measured at the highest carbon dioxide content compared to lower values. Obviously, not only carbonates or hydrogen carbonates play a role in the cover layer formation responsible for that phenomenon, but also the other salts present in the solution, e.g. sulfates in this case, as shown by the tests in oxygen-free distilled water – Figure 31. The consumption values can be directly assigned to the increasing carbon dioxide contents.

Figure 32 shows the corrosion rates determined for all 10 steel grades in 0.5 M sodium sulfate solution saturated with carbon dioxide. The corrosion rates of the unalloyed steel grades No. 1 to 4 were about 0.05 mm/a (1.97 mpy), with a low oxygen content in the carbon dioxide gas practically remaining without influence. The attack on the lightly alloyed steels No. 5 and 6, in contrast, was much more considerable. Obviously, these small contents of alloying elements contributing to the formation of cover layers in the atmosphere, which provide a better protection, lead to the formation of less protecting layers under these conditions.

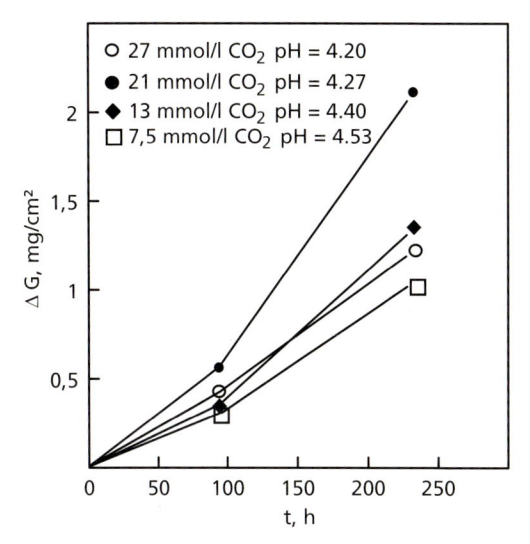

Figure 30: Time dependence of the material consumption rates of steel 1 in 0.5 M sodium sulfate solution with different oxygen-free CO_2 contents [72]

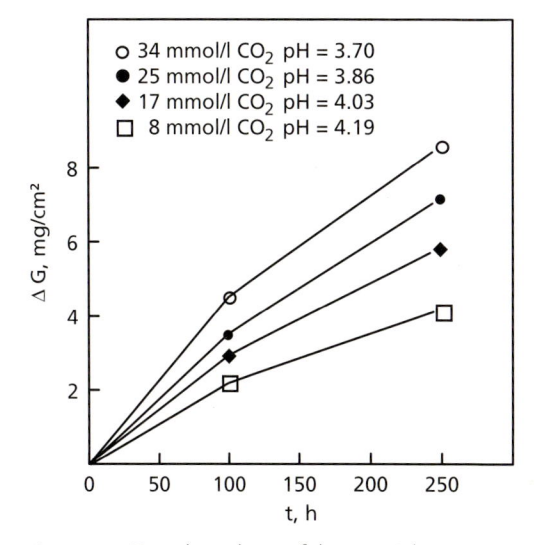

Figure 31: Time dependence of the material consumption rates of steel 1 in distilled water with different oxygen-free CO_2 contents [72]

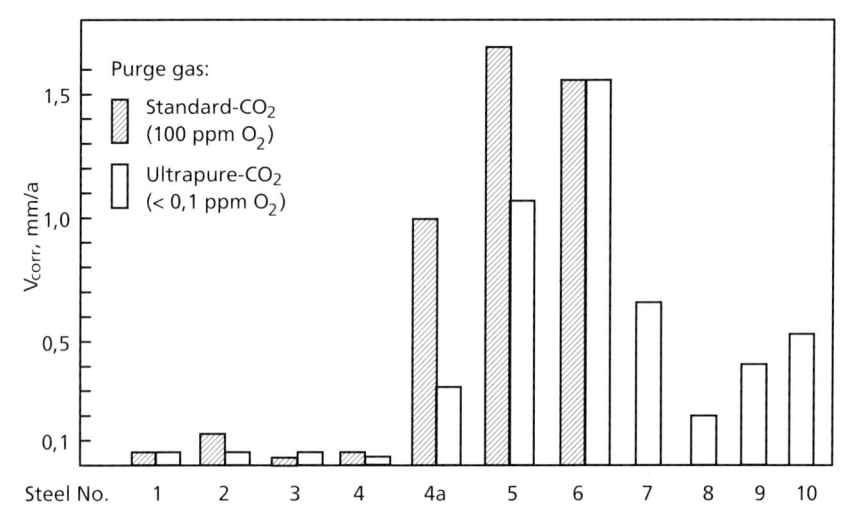

Figure 32: Corrosion rates of the 10 examined steels in 0.5 M sodium sulfate solution saturated with carbon dioxide at 298 K (25°C) [72]
(Steel 4a about 90% cold-formed)

To check which alloying elements in the steels No. 5 and 6 exert a negative influence on the corrosion behavior, the four special smelts No. 7 to 10 were examined, which were produced on the basis of pure iron with individual alloying elements, namely chromium, phosphorus and copper as well as copper plus phosphorus. It turned out that all three alloying elements reduced the corrosion resistance in carbonic solutions. Phosphorus exerts the rather lowest negative effect, yielding values for steel No. 8 which were only thrice as high as the values of the unalloyed steel grades No. 1 to 4. In contrast, the steels No. 7, 9 and 10 with increased copper and chromium contents exhibited corrosion rates which were up to ten times higher than those of the unalloyed steel grades No. 1 to 4. On these steels loose corrosion products were formed, in which considerable contents of copper were found.

The influence of cold forming was examined in steel No. 4, which was cold-formed by about 90%. The corrosion rates determined for these specimens were five times higher in the oxygen-free test solution und even 30 times higher in the oxygen-containing solution compared to the non-formed specimens.

Compared to the test solution containing sodium sulfate, oxygen-free distilled water saturated with carbon dioxide exhibited a remarkably higher aggressiveness. Under conditions, which were identical otherwise, corrosion rates of 0.4 mm/a (15.7 mpy) were determined for unalloyed steel in this fluid. The reason is the lower pH value and the higher carbon dioxide content of these solutions. Only above pH 4.2 the presence of lower oxygen amounts resulting from the charge of carbon dioxide containing 100 ppm oxygen caused an acceleration of corrosion.

However, indicating linear material consumption rates is only of limited importance to the evaluation of the resistance of steels in the chosen fluid. All examined materials exhibited pitting corrosion after induction periods of several hundred hours, the occurrence of which was only attributed to the presence of carbon dioxide and/or carbonic acid. Regarding the pit diameter, pit depth and pit density, pitting corrosion was gradually weaker in low-alloy steels. In oxygen-free distilled water saturated with carbon dioxide shorter induction periods (about 250 hours) of pitting corrosion were observed compared to sulfate-containing solutions. The presence of oxygen traces caused an extension of pitting corrosion induction periods to a maximum of 1000 hours, in particular for low-alloy steels. Local corrosion can also be explained by assuming active and passive regions on the metal surface.

Also newer investigations attempted to answer the question about the influence of chromium additions in steel on the corrosion in media containing carbon dioxide [80]. The test materials were pure iron (99.95 %) and three alloys of pure iron with 1 %, 3 % and 5 % chromium additions. The pure materials were chosen to avoid precipitations. All materials had a ferritic structure and the chromium was completely dissolved in the matrix even at 5 %. The test medium was a sodium chloride solution saturated with CO_2 and stirred with an NaCl concentration of 116 g/l and a pH value adjusted to 4.2, 5.0 or 6.0 by means of HCl or NaOH. The tests were performed at room temperature. Access of oxygen was excluded. After a test duration of 12 h, 24 h, 48 h and 72 h the material consumption of the specimens was determined.

The critical corrosion medium in the condensates for the internal corrosion of pipes in petrochemical plants are water, gases with acid reaction, such as carbon dioxide or hydrogen sulfide, and volatile organic acids occurring as byproducts of the hydrocarbons. For this reason 20 g/l acetate were added to the solution in several tests.

The material consumption measured for the specimens in the acetate-free NaCl solution with pH 4.2 is plotted against time in Figure 33. All materials tested show a

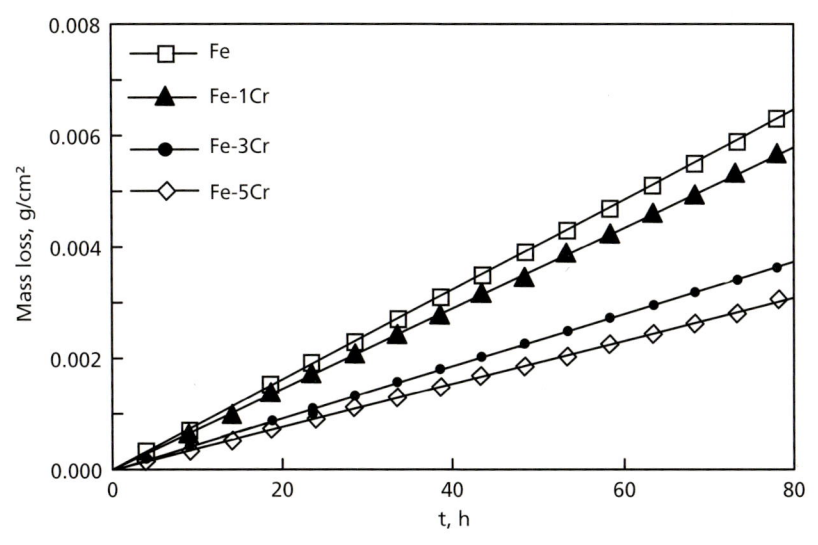

Figure 33: Material consumption of the four materials in acetate-free NaCl solution with pH 4.2 saturated with CO_2 [80]

linear time dependence of the material consumption. Although the corrosion values apparently decrease as the chromium content increases, the pure iron specimens and the specimens with 1 % Cr differ slightly only in their corrosion rates, whereas the behavior of the two higher alloyed materials is clearly better but again with slight differences only between them.

In Figure 34 the material consumption measured for the pure iron specimens and the specimens alloyed with 3 % Cr is plotted at different pH values.

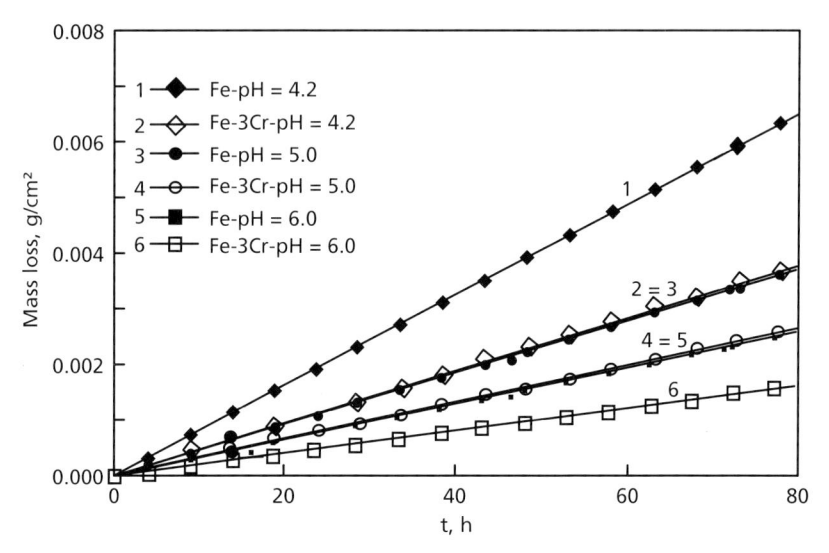

Figure 34: Material consumption of the pure iron specimens and the specimens alloyed with 3 % Cr in acetate-free NaCl solution saturated with CO_2 at different pH values [80]

Here again a linear time dependence of material consumption is found. The material consumption of both materials decreases as the pH value increases. A positive influence of the chromium content can only be detected at a low pH value.

Due to the linear dependence the corrosion rates indicated in Table 11 were extrapolated. According to the results indicated in [72] such linear time dependence can no longer be established for longer test durations (refer to Figure 29, Figure 30 and Figure 31).

Material		Fe	Fe-1Cr	Fe-3Cr	Fe-4Cr
	pH				
mm/a (mpy)	4.2	0.898 (35.4)	0.843 (33.2)	0.596 (23.5)	0.483 (19.0)
	5.0	0.52 (20.5)		0.395 (15.6)	
	6.0	0.34 (13,4)		0.342 (13.5)	

Table 11: Extrapolated corrosion rates of the examined materials in acetate-free NaCl solution saturated with CO_2 at different pH values [80]

The results of the tests during which acetate was added to the test solution as depicted in Figure 35 and Figure 36 suggest that there is no longer any linear time dependence of the material consumption under these test conditions.

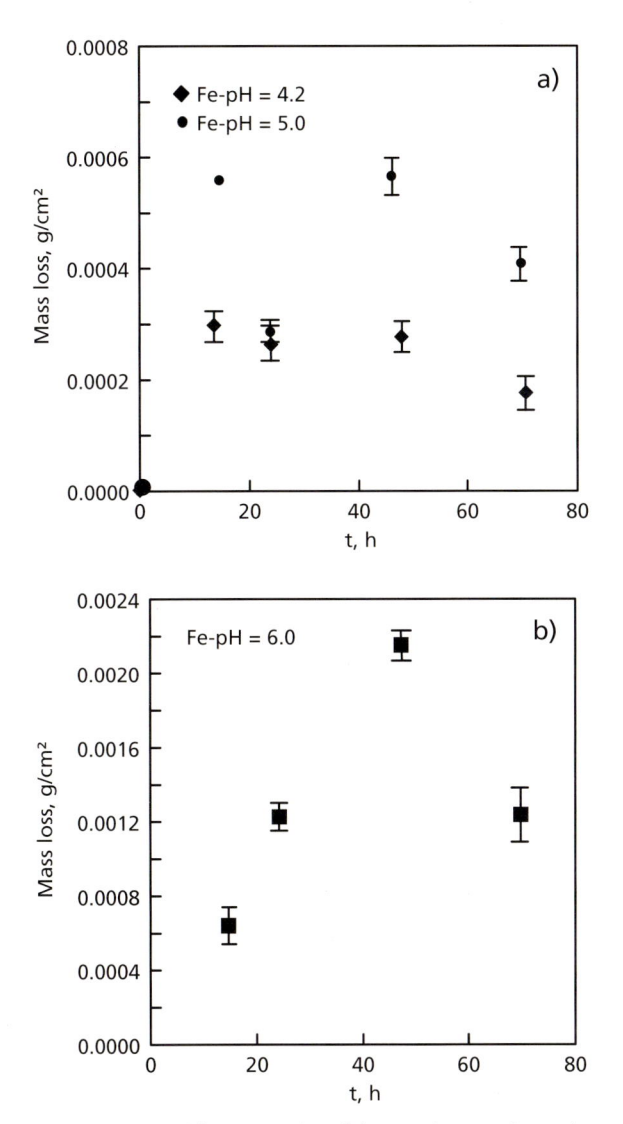

Figure 35: Material consumption of the pure iron specimens in acetate-containing NaCl solution saturated with CO_2 at different pH values [80]

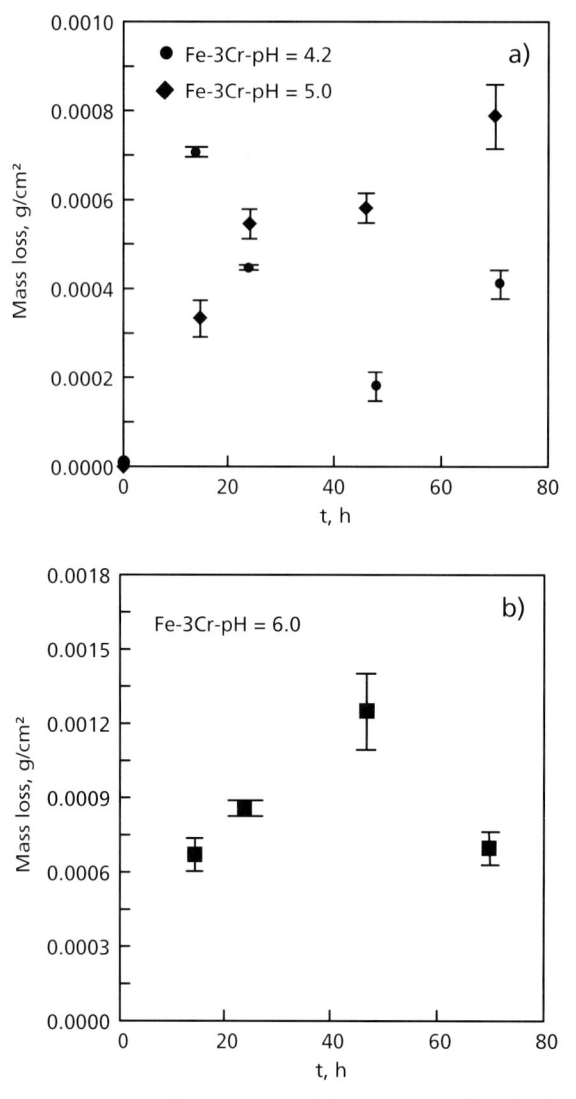

Figure 36: Material consumption of the specimens alloyed with 3 % Cr in acetate-containing NaCl solution saturated with with CO_2 at different pH values [80]

The influence of acetic acid on the carbon dioxide corrosion of unalloyed steels in the horizontal flow of two-phase water-steam mixtures was investigated in connection with the problems in oil and gas pipelines with different corrosion manifestations occurring in 12 and 6 o'clock positions [81, 82]. To this end, from a circulation apparatus with horizontally arranged pipes specimens of the steels X 65 (DIN Mat. No. 1.8975) and C 1018 were taken, the composition of which is indicated in Table 12, were laid out in the 12 o'clock position (gaseous phase) and in the 6 o'clock position (liquid phase) and checked for their corrosion resistance.

	C	Si	Mn	P	S	Cu	Cr	Ni	Al	Nb	V
X 65	0.130	0.260	1.160	0.009	0.009	0.131	0.140	0.360	0.032	0.017	0.047
C 1018	0.160	0.250	0.790	0.008	0.029	0.250	0.063	0.078	0.001	0.006	0.004

Table 12: Composition of the investigated steels [81]

The liquid phase of the test medium mainly consisted of an aqueous 1 % NaCl solution with different contents of acetic acid from 500 ppm to 5000 ppm and a pH value of 5.5. The gaseous phase consisted of a gas-water steam mixture. The partial pressure of CO_2 amounted to 1.54 bar and the total pressure to 2 bar. The test temperature was 353 K (80°C) and the flow rates were 0.1 m/s in the liquid phase and 2 m/s in the gaseous phase.

As shown in Figure 37 using the example of C 1018 steel specimens in the 6 o'clock position, the corrosion rates increase as the content of acetic acid in the solution increases. Irrespective of the acid content, the values first increase as the test duration increases and then drop to a constant value below 1 mm/a (39.4 mpy).

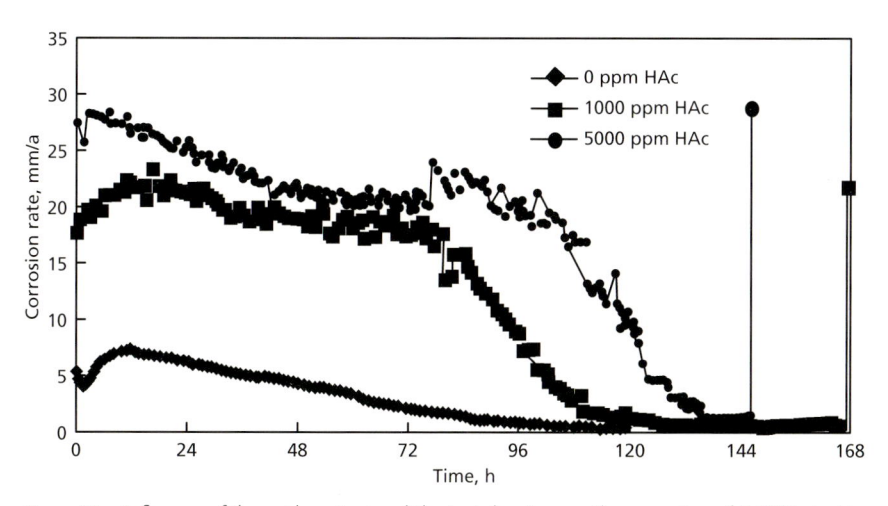

Figure 37: Influence of the acid content and the test duration on the corrosion of C 1018 steel in 1 % NaCl solution saturated with CO_2 with a pH value of 5.5 at 353 K (80°C) in the 6 o'clock position [81]

The specimens in the 12 o'clock position (gaseous phase) and the specimens of the X 65 steel basically follow the same corrosion behavior with a clear increase of the corrosion rates at an increasing acetic acid content in the solution. This is attributed to the contribution of the hydrogen ions from free acetic acid to the cathodic partial reaction of the corrosion process. Both in the 12 o'clock position and in the 6 o'clock position local deposits of $FeCO_3$ were formed, causing a local corrosive attack.

These local sites of attack can be found, in particular, in the upper pipe position, attributable to the enrichment of acetic acid in the condensate drops. This is also confirmed by other investigations in the circulation apparatus, which also present electrochemical examinations and model calculations proving the practical experience [83–86].

In this regard Figure 38 shows results of electrochemical investigations of pipe steel X 65 in 0.01 M NaCl solution saturated with 1 bar CO_2 and acetic acid additions of 0 to 1000 ppm at room temperature [86]. Up to an acetic acid content of about 300 ppm the corrosion potential rapidly shifts into the negative direction. In contrast, higher contents have a less strong influence.

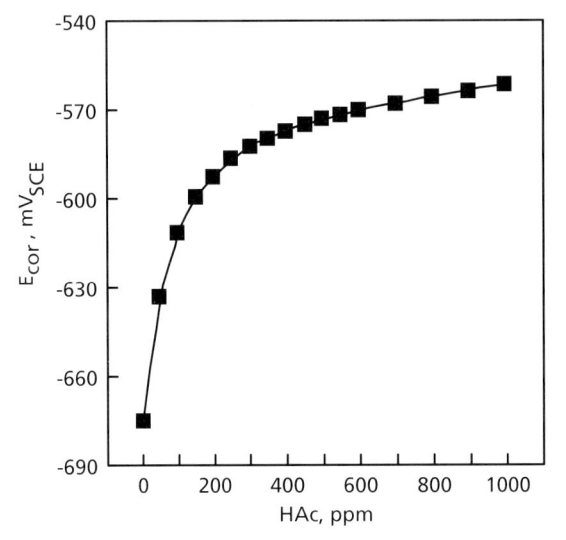

Figure 38: Corrosion potential of steel X 65 in 0.01 M NaCl solution saturated with 1 bar CO_2 and acetic acid additions at room temperature [86].

To avoid corrosion problems caused by aqueous phases containing carbon dioxide in the crude oil and natural gas production inhibitors on the basis of organic compounds with a higher molecular weight can be used.

Hydrogen-induced stress corrosion cracking was not detected in the fairly soft steels examined in [72]. However, features of anodic stress corrosion cracking were found at higher partial pressures of CO_2 (equal/above 10 bar) and higher temperatures (> 313 K (> 40°C)) if the steels were subjected to static tensile stresses above yield strength.

Stress corrosion cracking of heat treatable steels in CO_2-containing attacking media has long been known for the $CO_2/H_2S/H_2O$ system. In this medium CO_2 plays a rather contributing instead of a primary role in the corrosion process. However, it is also known that higher strength steels, in particular heat treatable steels, may be affected by stress corrosion cracking in wet carbon dioxide and in general in the $CO_2/CO/H_2O$ system.

But also in the simple two-substance CO_2/H_2O system stress corrosion cracking can occur. This turned out in damage investigations of numerous cracked or ruptured CO_2 gas bottles or fire extinguishers attributable to the residual moisture resulting from a previous water pressure test. Here, the crack path was usually an intercrystalline one. From all experience it can be concluded that stress corrosion cracking depends on the presence of CO_2 and CO as well as water. No damage is known if only CO_2 or only CO were present. Partial pressures of 60 mbar of any gas type turned out to be sufficient to cause stress corrosion cracking [87].

In systematic investigations with heat treatable low-alloy steels of grade 36Mn7 (DIN Mat. No. 1.5069; old: 37Mn5), 34CrMo4 (DIN Mat. No. 1.7220) and 30CrNiMo8 (DIN Mat. No. 1.6580), which were present in three strength levels, it was found out in experiments at a constant force of 50 to 90 % of the yield strength as well as at a constant strain rate between 10^{-5} and 10^{-7} 1/s in the temperature range from 273 to 333 K (0 to 60°C) and at a partial pressure of CO_2 from 1 to 60 bar that all examined materials may suffer stress corrosion cracking even at constant load [88]. The results are summarized in Figure 39.

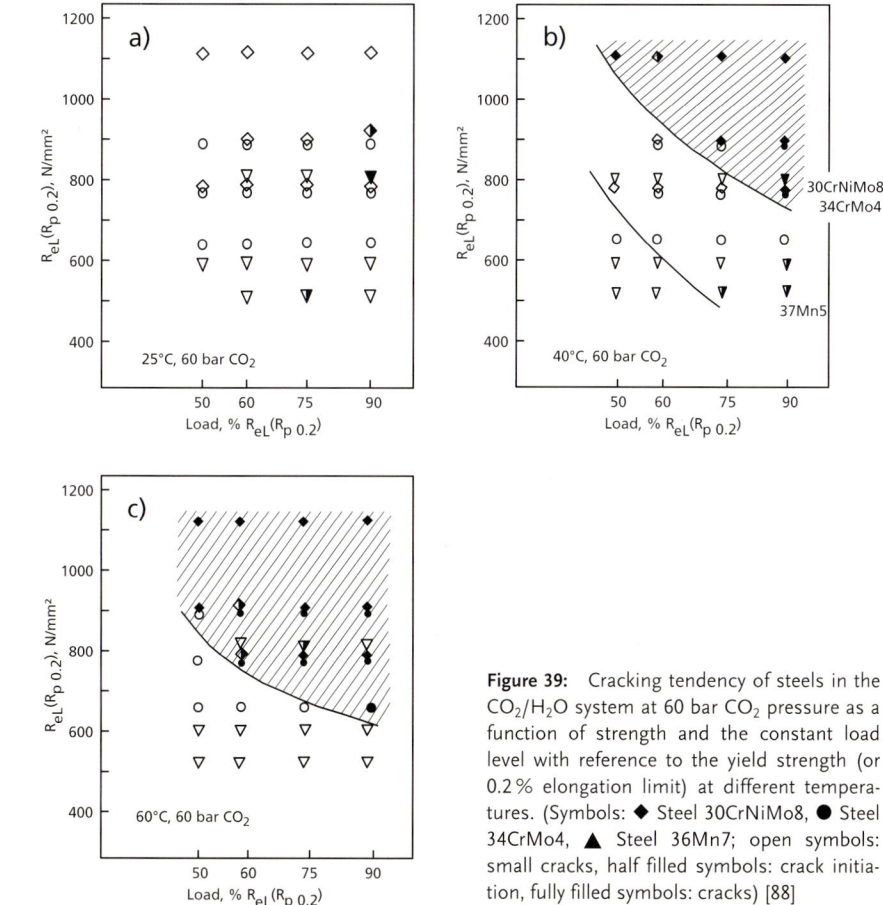

Figure 39: Cracking tendency of steels in the CO_2/H_2O system at 60 bar CO_2 pressure as a function of strength and the constant load level with reference to the yield strength (or 0.2 % elongation limit) at different temperatures. (Symbols: ◆ Steel 30CrNiMo8, ● Steel 34CrMo4, ▲ Steel 36Mn7; open symbols: small cracks, half filled symbols: crack initiation, fully filled symbols: cracks) [88]

Here, the main influencing factors are the strength of the steel, the level of loading, the partial pressure of CO_2 and the temperature. The cracking tendency rises as the strength of steel, the loading level and the temperature increase. At 333 K (60°C) and 60 bar partial pressure of CO_2, for instance, the region can be defined with high precision for all examined steels in which stress corrosion cracking occurs depending on the material strength and the level of loading. For the low-alloy Cr steels this is also possible at 313 K (40°C), the critical region being shifted towards higher strengths. The propensity to stress corrosion cracking is especially high at high partial pressures of CO_2. Above 10 bar CO_2 the lifetimes of constantly loaded tensile test specimen (0.9 Re) was generally below 1000 h.

However, at this temperature the steel 36Mn7 exhibited a different behavior. Although at loads < 0.9 R_e cracks propagating from the surface were no longer found, crack initiation occurred in the V-notch shaped pitting corrosion sites, preferably developed at the breakthrough point of impact lines. The formation of crack initiation sites can be explained by the much higher stress intensity at the notch root. With reference to the real stress conditions, the susceptibility range for stress corrosion cracking indicated for the low-alloy Cr steels would also apply to the steel 36Mn7.

However, the transcrystalline damage caused by stress corrosion cracking on the outer side of gas transport lines with PE jacketings is indicative of the possibility that stress corrosion cracking in CO_2-containing aqueous media is also possible at low partial pressures of CO_2. The time until a crack is initiated may then cover a period of several years.

For all examined steel grades also mass loss measurements were performed in a temperature range from 273 to 353 K (0°C to 80°C) and at partial pressures of CO_2 from 1 bar to 60 bar. The results are summarized in Figure 40.

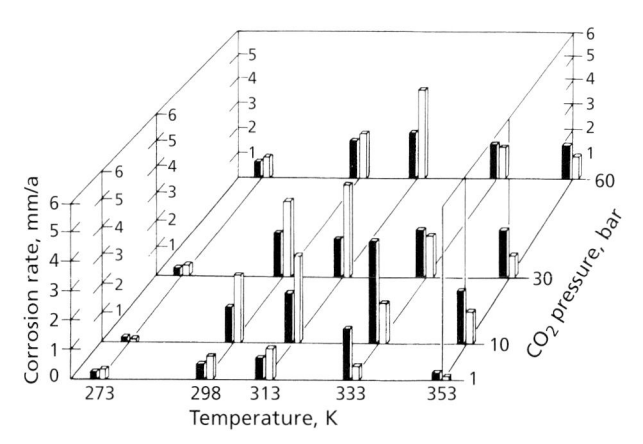

Figure 40: Influence of temperature and CO_2 pressure on the corrosion of heat treatable steels in the CO_2/H_2O system [88]

The corrosion rates increase with increasing temperature and carbon dioxide pressure. At 313 K (40°C) the steels 30CrNiMo8 and 34CrMo4 are less attacked than 36Mn7 steel, whereas the attack is stronger than that at higher temperatures. At pressures up to 30 bar a maximum of uniform surface corrosion can be observed at about 333 K (60°C).

It can be seen also from these results that the protective action and stability of the carbonate cover layers are a function of temperature and CO_2 pressure. In general, the relative protective effects of the cover layers formed at temperatures above 313 K (40°C) was better the higher the CO_2 pressure was. The cover layers formed under these conditions also exhibited a much poorer dissolution behavior when being stripped compared to the corrosion products grown at lower temperatures and lower CO_2 pressures. A comparison of the results obtained from material consumption measurements with the results of the tensile tests shows that the highest susceptibility to stress corrosion cracking was found under conditions where also the highest pitting corrosion tendency was found. This is reasonable since local corrosion can only occur under the conditions of instable cover layers.

Regarding the question about the mechanism of the crack-forming corrosion in the CO_2/H_2O system, all observations are indicative of the fact that it is a classical anodic stress corrosion cracking, the formed hydrogen supporting crack propagation. The assumption of an anodic stress corrosion cracking is further supported by the fact that the cracking tendency increases as the temperature increases. In case of a pure hydrogen-induced stress corrosion cracking the highest susceptibility would rather occur at lower temperatures (around 298 K (25°C)) than at higher temperatures as found.

Scanning electron microscopic examinations of fracture surfaces and forced ruptures of tensile test specimens damaged by stress corrosion cracking did not reveal any impact fracture morphologies referring directly to hydrogen-induced crack formation. On the other hand, considerable hydrogen permeation rates were found in hydrogen permeation measurements, in particular at an elevated temperature as well as under the influence of tensile stress.

In the 60s and 70s crack damage caused by transcrystalline stress corrosion cracking was found on gas containers, bottles and separators of city and syngas plants of unalloyed and low-alloy steels if $CO/CO_2/H_2O$ mixtures were applied under operating conditions [89].

To date transcrystalline damage of a similar kind has been preferably found in the presence of a hydrogen-induced crack mechanism, e.g. in H_2S-containing sour gas condensate. Due to different signs and indications, the investigations performed did not permit a clear decision as to which mechanism of stress corrosion cracking was responsible here. Further investigations aimed at clarifying the damage mechanism and examine the influencing factors of media and materials.

For the investigations the fine-grain structural steels S355N (DIN Mat. No. 1.0545; old: StE 355) and S500N (DIN Mat. No. 1.8907; old: StE 500) were chosen as well as the steels S235JR (DIN Mat. No. 1.0037; old: St 37-2) and 13CrMo4-5 (DIN Mat. No. 1.7335; old: 13 CrMo 4 4). The test sheets were provided in a normalized condition. Fine-grain structural steel S355N is widely used for the manufacture of vessels and apparatuses in the chemical industry and was, therefore, used for these investigations as the basic material. The other steels mentioned served for comparative investigations.

Round test bars of the steels were tested in autoclaves in an electrochemically controlled process, with CO/CO_2-containing water (partial pressures up to 20 bar CO_2 and 10 bar CO) being applied with various potentials, temperatures and oxygen contents.

The investigation was mainly performed as slow strain rate tensile tests (CERT tests) during which the specimens were strained until fracture (in most cases at strain rates of 4.3×10^{-5}/s and 8.6×10^{-6}/s). Additional examinations at constant load (125 % R_{eL}) provided further information about the mechanism of crack propagation.

A transcrystalline crack formation occurred on all steel grades in $CO/CO_2/H_2O$ mixtures both under cathodic and anodic polarization. Regarding the free corrosion potential, the sensitivity to cracking was neglectably small. Crack formation is intensified in all materials examined on both increasing anodic and increasing cathodic polarization.

At a slower strain rate, the depth of cracks is clearly higher. Instead of polarization with direct current, the adjustment of the free corrosion potential in the range of stress corrosion cracking sensitivity can be also reached by the addition of oxygen (air).

Only anodic stress corrosion cracking can be the cause of the damage in practice since the cathodically induced crack propagation is not typical for the $CO/CO_2/H_2O$ system but is observed in all weakly acidic hydrogen-supplying media and is of no relevance to practice as such conditions (cathodic current; rapid elongation) normally do not occur. The presence of atmospheric oxygen is responsible for the adjustment of the critical potentials.

In the potential range of anodic stress corrosion cracking the crack formation increases with increasing temperatures (up to about 333 K (60°C)), whereas surface corrosion increases at temperatures above that value and exclusively occurs at a temperature above 373 K (100°C).

Even under static load conditions (125 % R_{eL}) all steel specimens cracked in $CO/CO_2/H_2O$ mixtures on anodic polarization (lifetimes < 20 h).

The localization of the attack (crack nucleation) is also attributable to the formation of cover layers, which can be observed in the corrosion test in that the material consumption rates are considerably reduced (by about one order of magnitude).

Without polarization but with addition of oxygen (partial pressure of 2 bar) all specimens also failed, with the steels S355N and 13CrMo4-4 exhibiting especially short lifetimes. Only in air-flushed water the steel S355N did no longer exhibited any cracks.

S235JR steel reached a lifetime of up to 70 h, whereas the lifetime of the 13CrMo4-4 steel remained below 10 h. Due to its tendency to form cover layers, 13CrMo4-4 steel responds already to very small additions of oxygen.

The gas composition may vary over wide ranges without loosing effect. Only regarding the CO_2 content a lower limit concentration is required for crack nucleation. At CO_2 contents below 1 bar the crack nucleation is rather weak.

Therefore, anodic stress corrosion cracking can be considered the cause of the damage occurred in practice. An anodic polarization, which is finally the cause of a stress corrosion cracking risk, is easily reached by the presence of oxygen or other oxidants. At a moderately increased temperature (333 K (60°C)) the crack intensity of anodic stress corrosion cracking increases.

All tested materials are susceptible to cracking and exhibit only little difference in their sensitivity. However, the material 13CrMo4-4 is especially sensitive to small amounts of oxygen. Its alloying elements facilitate the formation of a cover layer with a strong protective effect resulting in a marked localization of the attack and a rapid crack growth.

A 17 Ferritic chrome steels with < 13 % Cr

In carbon dioxide at elevated pressures and temperatures the ferritic heat resisting chromium molybdenum steels with chromium contents of 9 % and 12 % exhibited an accelerated oxidation following a local scale nodule similar to the unalloyed and low-alloy steels. Below temperatures of 823 K (550°C) local breakthroughs do not occur. Above this temperature the oxidation process can be described again by means of the curve schematically depicted in Figure 16 comprising three phases. In the first phase, during which protecting oxide layers are formed, higher silicon contents up to 1 % in the steel exert a favorable effect. A decreasing water vapor content in carbon dioxide also reduces the oxidation rate. The first break-offs in the oxide layer occur at a mass increase of 20 mg/cm^2. The time until breakthroughs occur is reduced by increasing contents of carbon dioxide in water vapor and carbon monoxide. Higher silicon contents have practically no influence on the oxidation behavior in the third phase.

As in the low-alloy steels the protecting oxide layer has again a two-layer structure with a compact outer magnetite layer and a porous inner spinel layer enriched with chromium and silicon as oxides. And again the oxide layers contain higher amounts of carbon with a maximum amount at the steel/scale phase boundary. Also the base material is considerably carburized.

Behavior in aqueous carbon dioxide solutions

In aqueous solution of carbon dioxide with a partial pressure of 13.8 bar and a temperature of 328 K (55°C) the corrosion rate of chromium alloyed steels decreases if the chromium content increases as shown in Table 13 [90].

Chromium content %	2.25	5.0	9.0	12.0
Corrosion rate mm/a (mpy)	1.5 (59.1)	1.17 (46.1)	0.04 (1.57)	0.02 (0.79)

Table 13: Corrosion rate of chromium alloyed steels in CO_2 containing water after a test duration of 7 days [90]

With an increasing test duration the corrosion rates further decrease as a result of cover layer formation and in case of steel with 2.25 % Cr drop to

– 0.16 mm/a (6.30 mpy) after 28 days,
– 0.08 mm/a (3.15 mpy) after 70 days

and in case of steel with 12.0 % Cr drop to

– 0.005 mm/a (0.20 mpy) after 70 days.

A 18 Ferritic chrome steels with ≥13 % Cr
A 19 High-alloy multiphase steels
A 19.1 Ferritic/perlitic-martensitic steels

As shown in Table 13 the steels with higher chromium contents are largely resistant to general corrosion in carbon dioxide-containing waters. But in aqueous solutions containing also hydrogen sulfide apart from carbon dioxide, as for instance occurring in the aqueous phases during the production and the transport of natural gas and crude oil, the chromium alloyed steels may be subject to the risk of stress corrosion cracking.

By reducing the carbon content and adding the alloying elements nickel, molybdenum, copper and nitrogen, the corrosion behavior of a common 13 % chrome steel according to API in CO_2 and H_2S-containing media can be clearly improved [91]. Table 14 shows the chemical analytical values of a modified steel compared to 13 % chrome steel.

Steel	C	Si	Mn	P	S	Cr	Ni	Cu	Mo	N
CRSS	0.017	0.30	0.51	0.017	0.001	12.8	5.90	1.56	1.97	0.017
API	0.190	0.25	0.58	0.011	0.003	12.6	0.13			

Table 14: Chemical compositions (% by weight) of both 13 % chromium steels examined [91]

Tests were performed to investigate the general corrosion and the stress corrosion cracking using four-point bending specimens and tensile test specimens. A load of 100 % of the yield strength was applied as a test stress to tensile test specimen and a load of 90 % of the yield point was applied to the bending test specimens. Following the NACE Specification TM0177-90A the test duration was 720 hours each. Solutions containing acidic chloride, hydrogen sulfide and carbon dioxide as occurring in plants processing natural gas and crude oil were chosen as the test solutions. The test conditions and results are compiled in Table 15 and Table 16.

The higher resistance to general corrosion compared to normal 13% chrome steel was attributed to the additional alloy contents of nickel and copper, whereas the better resistance to stress corrosion cracking was attributed to the molybdenum content.

Test solution	20% NaCl + 100 mg/l NaHCO$_3$ 40 bar CO$_2$ / 0.1 bar H$_2$S Temperature: 433 K (160°C); pH value: 4.1	
	CRSS	API
Bending test specimens	no cracks	no cracks
Corrosion rate	0.027 mm/a (1.06 mpy)	0.30 mm/a (11.8 mpy) *)

*) Local attack

Table 15: Test conditions and results for general corrosion and stress corrosion cracking of bending test specimens [91]

Test solution					Steel	
Chlorides ppm	CO$_2$ bar	H$_2$S bar	pH		CRSS	API
7200	50	0.02	2.9		no cracks	cracks
55800	50	0.02	2.7		cracks	cracks
1000	45	0.10	3.0		no cracks	cracks
68000	45	0.10	3.3		no cracks	cracks
68000	20	0.10	3.8		cracks	cracks

Table 16: Test solutions and results of stress corrosion cracking tests with tensile test specimens [91]

The good resistance to stress corrosion cracking of a steel containing 18% chromium and 2% molybdenum is also confirmed in [92].

Apart from the resistance to general corrosion, pitting corrosion and stress corrosion cracking, frequently materials used in the production of gas and oil are also required to be resistant to erosion corrosion. In this connection tests were performed with X20Cr13 steel to establish the erosion corrosion behavior in simulated carbon dioxide-containing formation water at 333 K (60°C) [93, 94].

Composition of the test solution: 3.80 % NaCl
0.44 % $CaCl_2$
0.07 % $MgCl_2$
partial pressure of CO_2: 3 bar
pH value: 4.2.

As a solid body 0.1 % sand with a grain size of 0.4 mm was added. A segmented coiled column served as the test track as schematically shown in Figure 41. The figure contains the material removal obtained plotted against the various Reynolds numbers Re for the X20Cr13 steel. For comparison, the values obtained for the unalloyed steel C15 at Re = 3.5 · 10^5 are also indicated. At the beginning of the pipe construction the maximum material removal is reached by working off the edges. Following the narrow pipe section a maximum is found again between 2 and 3 D_0^{-1}. Both maximum values depend on the flow rate. According to these investigation results a flow rate with a Reynolds number of 2.5 · 10^5 and a corrosion rate of about 0.5 mm/a (19.7 mpy) would still be acceptable for technical applications.

Compared to an unalloyed steel C15 the erosion corrosion resistance of the X20Cr13 steel in carbonated aqueous solution is clearly better. The erosion corrosion behavior of steel X20Cr13 was examined in the same test medium with the Reynolds number 2.5 · 10^5, however with different solid contents. The results are shown in Figure 42.

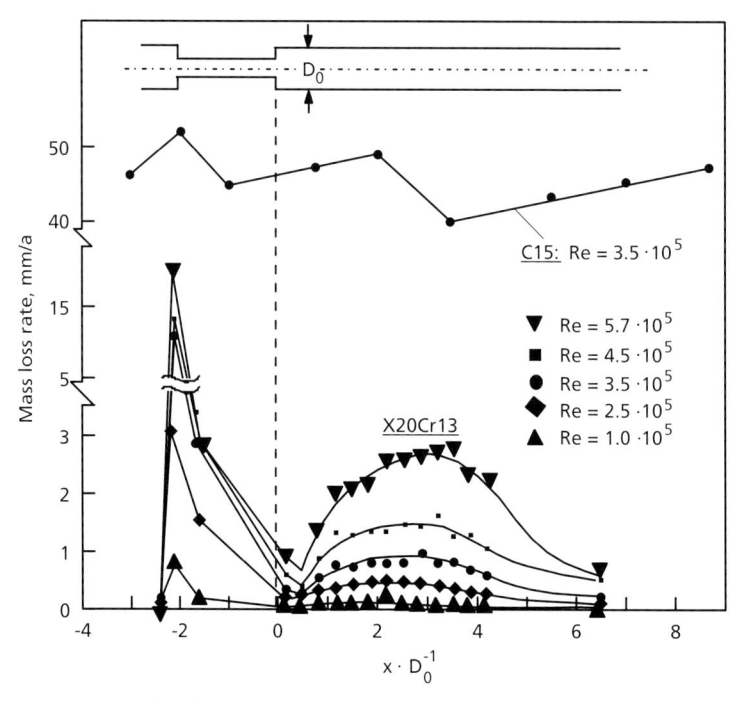

Figure 41: Results of erosion corrosion tests in carbonated water with different additions of solids [95, 96]

An increase in the solid particle content of the corrosion medium leads to an increase in the material removal, resulting in an almost linear dependence between the sand content and the erosion corrosion rate in the carbonated water. Reduction of the chloride content in the test solution remained without any significant influence on the material removal. In the temperature range from 298 to 333 K (25 to 60°C) a change in the erosion corrosion loss cannot be detected. Increasing the temperature from 333 to 353 K (60 to 80°C) leads to an increase in the corrosion rate by 25 % [93, 94].

Using steel X2CrNiMoN22-5-3 instead of X20Cr13 may reduce corrosion by another 50 %.

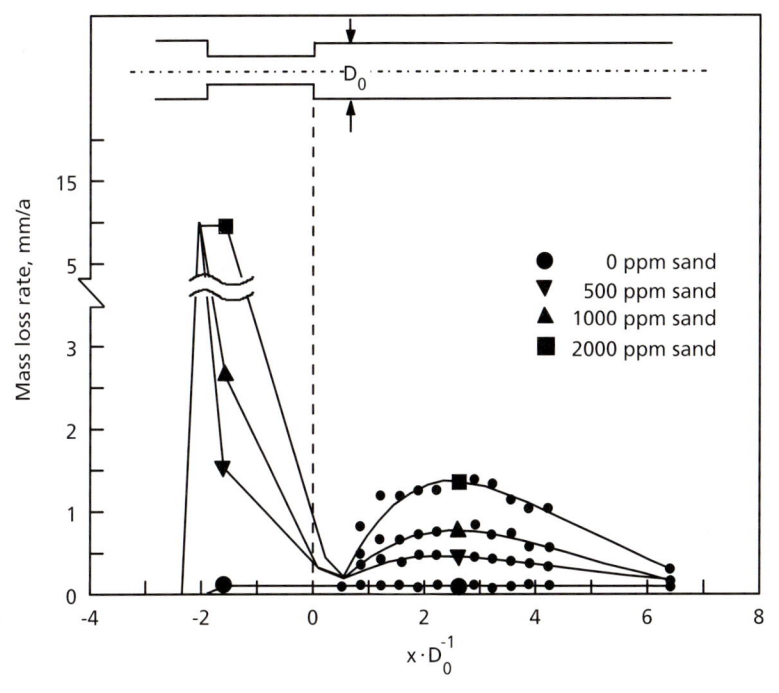

Figure 42: Results of erosion corrosion tests of steel X20Cr13 in carbonated waters with different solid additions [96]

The corrosion resistance of a 13 % Cr steel against carbon dioxide-containing waters may be increased by adding copper, nickel and molybdenum [97]. Starting from a common 13 % Cr steel six steel variants were investigated, the chemical composition of which is indicated in Table 17.

C	Si	Mn	P	S	Cr	Cu	Ni	Mo
0.025	0.25	0.45	0.015	0.002	13	0 to 2.0	4.0 to 5.0	1.0 to 2.0

Table 17: Chemical composition of the investigated steels [97]

The tests to investigate the behavior in carbon dioxide-containing solutions were performed by exposing the specimens in the autoclave to a 20% NaCl solution saturated with 3 MPa CO_2 gas at 453 K (180°C). The corrosion rate was calculated from the material consumption of the specimens after a test duration of 7 days. The results are shown in Figure 43.

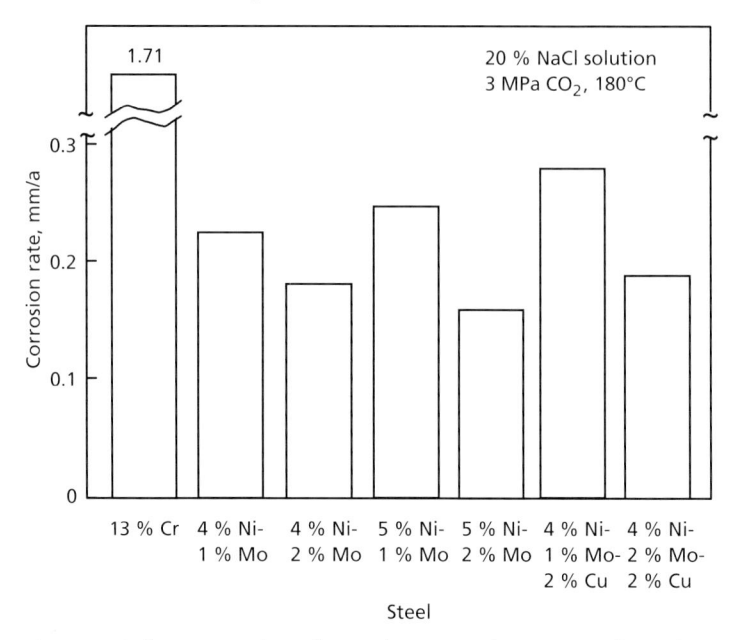

Figure 43: Influence on various alloying elements on the corrosion of a 13% Cr steel against carbon dioxide-containing solution at 453 K (180°C) [97]

The addition of the alloying elements mentioned leads to a clear reduction of the corrosion rates compared to non-modified 13% Cr steel. At the best, i.e. steel with 5% Ni and 2% Mo, the corrosion rate is reduced from 1.7 mm/a (66.9 mpy) to about 0.15 mm/a (5.91 mpy). Also the behavior with regard to stress corrosion cracking if carbon dioxide and hydrogen sulfide are present at the same time, is improved by the addition of alloying elements when comparing modified steels and conventional 13% Cr steel, the major effect being attributed to molybdenum.

A 20 Austenitic chromium-nickel steels

Stainless austenitic steels are widely used in power plants as well as in chemical and petrochemical plants in gas atmospheres exerting an oxidizing or carburizing effect, such as air, steam, carbon dioxide or combustion gases in a temperature range from 573 to 973 K (300°C to 700°C). The resistance to aggressive gases at high temperatures is the result of the formation of a chromium-rich cover layer. Therefore, the steels used under these conditions contain not less than 20% chromium.

The publications described in [98, 99] deal with the resistance of stainless steels with 20 % Cr and 25 % Ni in CO_2/CO gas mixtures. The chemical composition of the three investigated steels is shown in Table 18.

Steel	C	Si	Mn	P	p	Cr	Ni	Ti	Nb
20Cr25Ni/Nb	0.02	0.56	0.70		0.007	20.4	25.1		0.60
20Cr25Ni/Ti	0.015	0.66	0.81	0.005	0.001	19.7	24.1	0.23	
20Cr25Ni/TiNb	0.01	0.86	0.65			20.1	24.8	1.70	0.12

Table 18: Chemical composition of the three investigated steels [98]

The specimens with the dimensions of 20 mm × 5 mm × 0.4 mm from both steels 20Cr25Ni/Nb and 20Cr25Ni/Ti were subjected to recrystallization annealing in hydrogen at 1203 K (930°C) for 1 h, yielding grain sizes of ~ 9 μm and ~ 2.5 μm in the surface area and of 13 μm and 20 μm in the core area. The third alloy 20Cr25Ni/TiNb with 1.7 % Ti was nitrided for 1 h at 1423 K (1150°C) in an atmosphere consisting of 95 % nitrogen and 5 % hydrogen, a finely dispersed TiN phase forming in the steel matrix. The mean grain size in the core was around ~ 100 μm and on the specimen surface about ~ 1 μm.

The tests were performed in carbon dioxide with different CO contents of 2 %, 10 %, 30 %, 75 % as well as in pure carbon monoxide at 923 K, 1023 K and 1123 K (650°C, 750°C and 850°C). The gas phase equilibria of both reactions according to

Equation 14 $2\,CO \leftrightarrow CO_2 + C$

Equation 15 $2\,CO + O_2 \leftrightarrow 2\,CO_2$

were used to calculate the partial pressures of oxygen P_{O_2} and the carbon activities a_c for the various partial CO pressures and test temperatures as indicated in Table 19 according to Equation 8 and Equation 10.

Partial CO pressure bar	Temperature 923 K (650°C)		Temperature 1023 K (750°C)		Temperature 1123 K (850°C)	
	a_c	P_{O_2}/bar	a_c	P_{O_2}/bar	a_c	P_{O_2}/bar
0.02	1.29×10^{-3}	3.83×10^{-19}	1.46×10^{-4}	5.38×10^{-18}	2.45×10^{-5}	5.16×10^{-17}
0.10	3.51×10^{-2}	1.29×10^{-20}	3.97×10^{-3}	1.82×10^{-19}	6.67×10^{-4}	1.74×10^{-18}
0.30	4.07×10^{-1}	8.69×10^{-22}	4.59×10^{-2}	1.22×10^{-20}	7.73×10^{-3}	1.17×10^{-19}
0.75	7.12	1.77×10^{-23}	8.04×10^{-1}	2.49×10^{-22}	1.35×10^{-1}	2.34×10^{-21}
0.9998	1.58×10^{4}	6.39×10^{-30}	1.79×10^{3}	8.97×10^{-29}	3.00×10^{2}	8.60×10^{-28}

Table 19: Calculated carbon activities a_c and partial oxygen pressures P_{O_2} for various CO contents and test temperatures [98]

The mass increase of the specimens as a function of the exposure duration and the CO content of the test gas is shown for the three steels in Figure 44, Figure 45 and Figure 46.

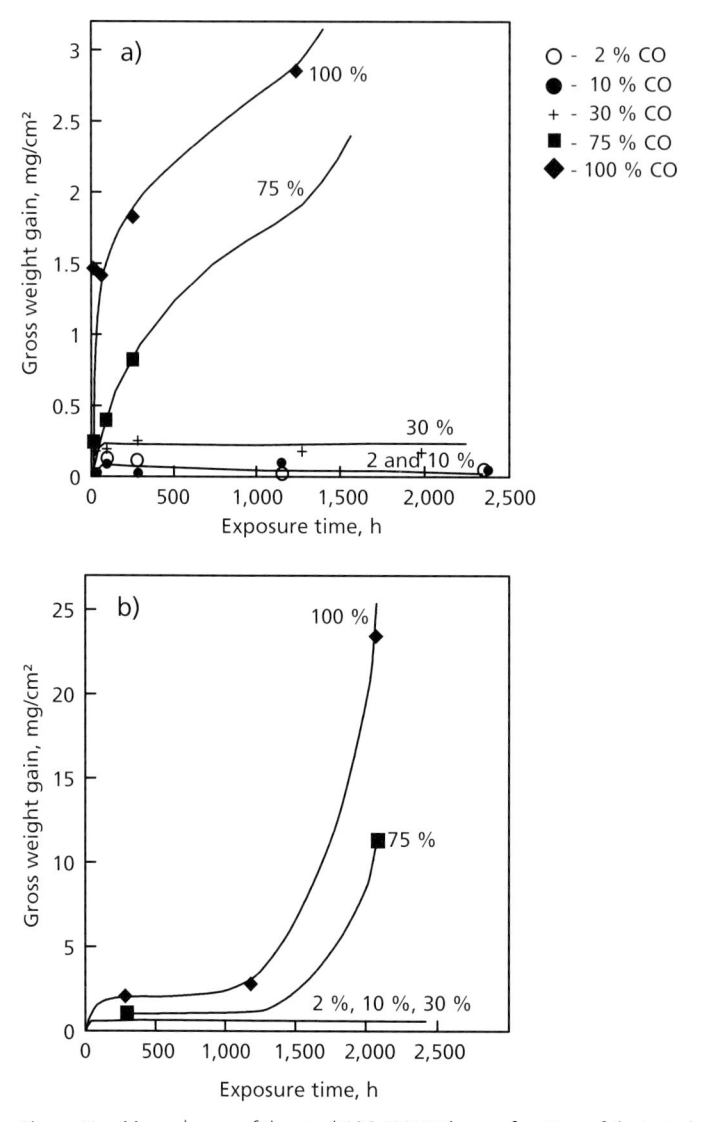

Figure 44: Mass change of the steel 20Cr25Ni/Nb as a function of the test duration and the CO concentration at 923 K (650°C) [98]

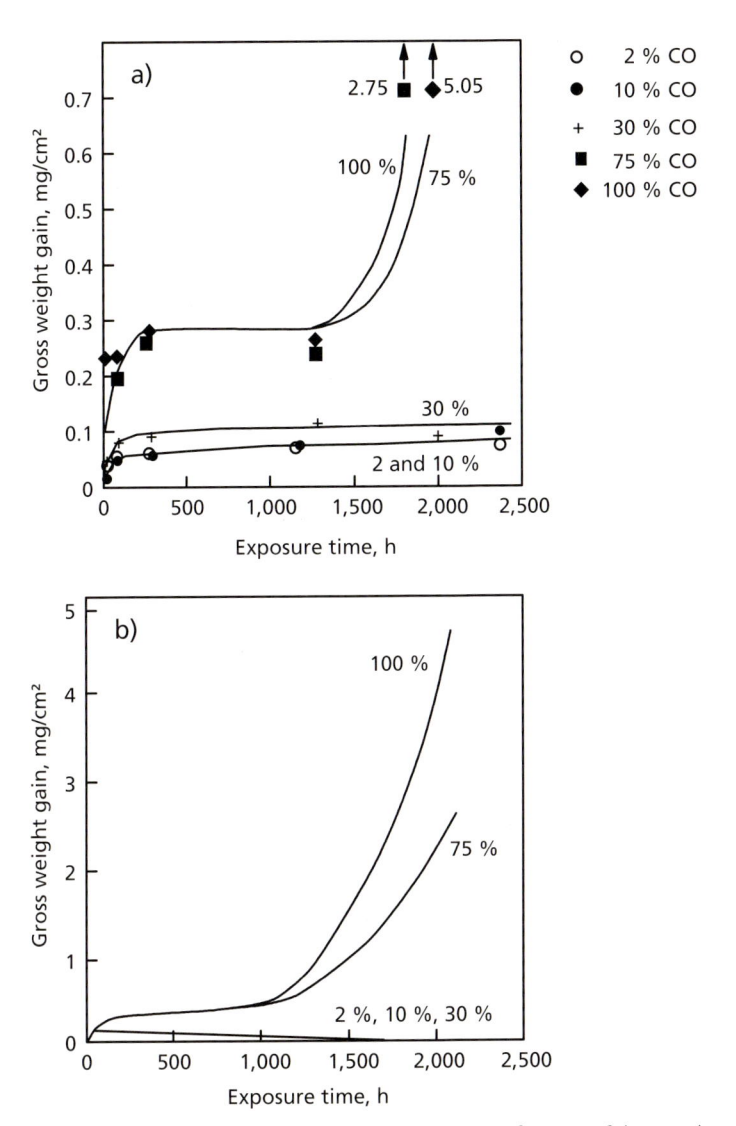

Figure 45: Mass change of the steel 20Cr25Ni/Ti as a function of the test duration and the CO concentration at 923 K (650°C) [98]

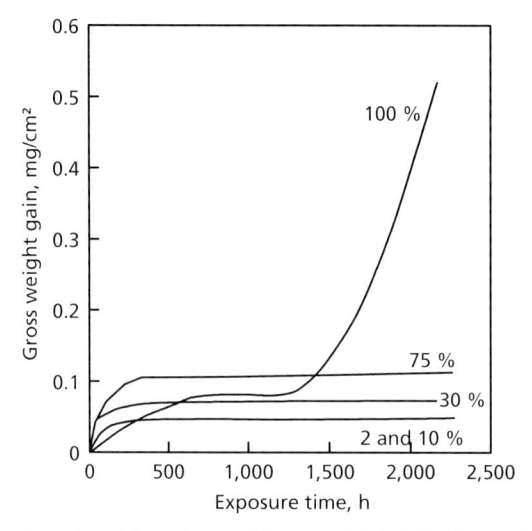

Figure 46: Mass change of the steel 20Cr25Ni/TiNb as a function of the test duration and the CO concentration at 923 K (650°C) [98]

Following a short rise in the mass increase all steels reach an almost constant value in the gases with CO contents up to 30 %. At higher CO contents a further rise is observed over a longer test duration, which is highest in case of 20Cr25Ni/Nb steel and occurs in pure carbon monoxide only in case of 20Cr25Ni/TiNb steel.

Figures 47 to 52 show the relevant curves of the mass increase as a function of the test duration and the gas composition for both test temperatures 1023 K (750°C) and 1123 K (850°C).

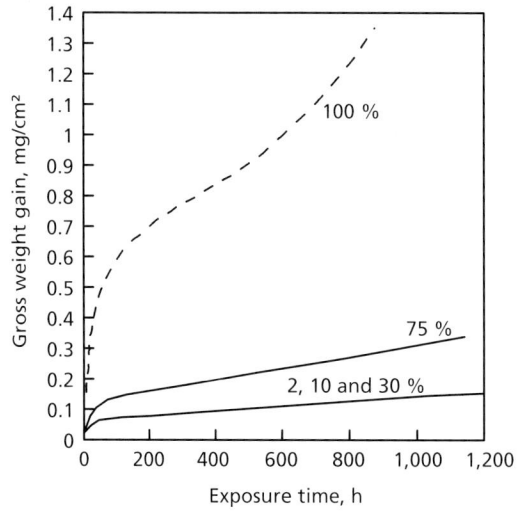

Figure 47: Mass change of the steel 20Cr25Ni/Nb as a function of the test duration and the CO concentration at 1023 K (750°C) [98]

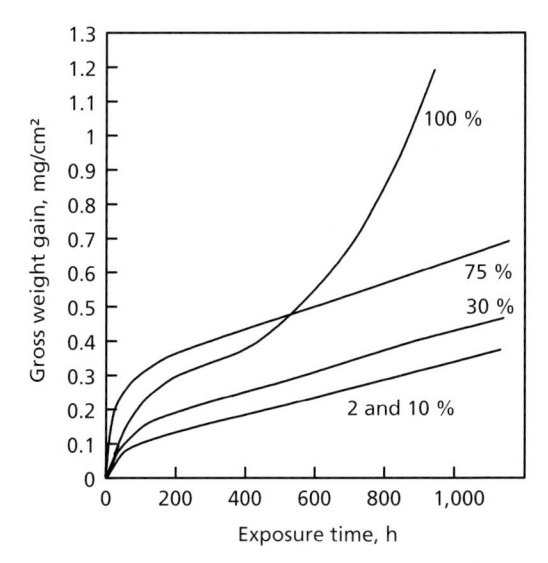

Figure 48: Mass change of 20Cr25Ni/Ti steel as a function of the test duration and the CO concentration at 1023 K (750°C) [98]

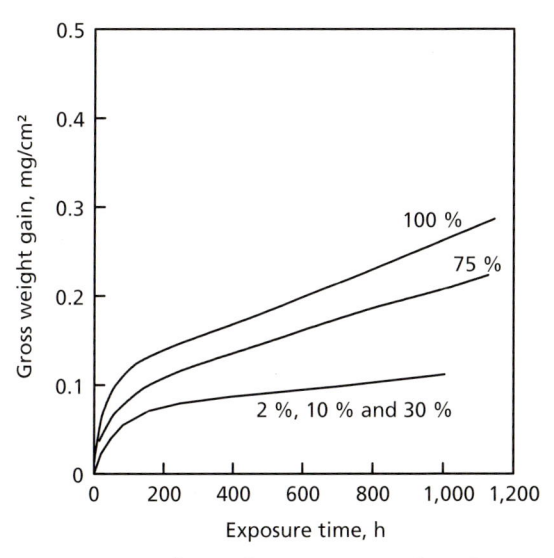

Figure 49: Mass change of 20Cr25Ni/TiN steel as a function of the test duration and the CO concentration at 1023 K (750°C) [98]

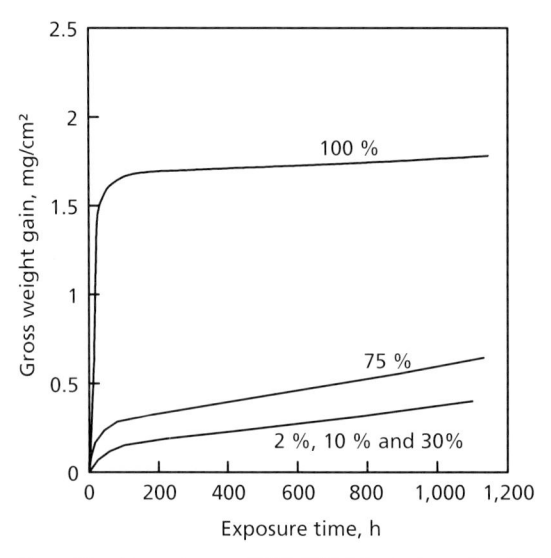

Figure 50: Mass change of 20Cr25Ni/Nb steel as a function of the test duration and the CO concentration at 1123 K (850°C) [98]

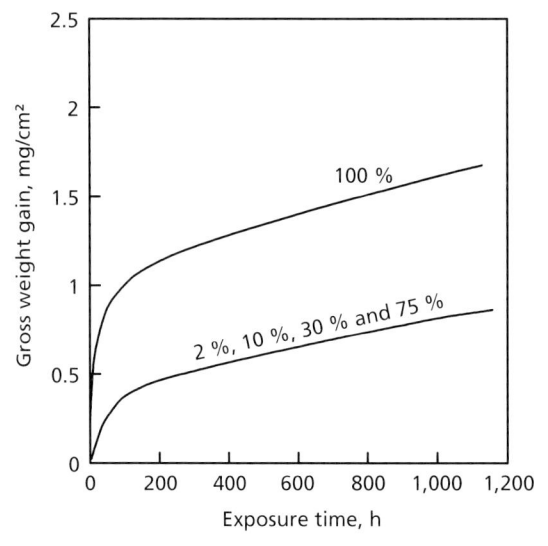

Figure 51: Mass change of 20Cr25Ni/Ti steel as a function of the test duration and the CO concentration at 1123 K (850°C) [98]

The curve at a test temperature of 1023 K (750°C) is almost identical with that at 923 K (650°C), however the TiN-containing steel does not exhibiting any rapid rise of the mass increase in pure carbon monoxide at longer test durations either. A rise of the mass increase in pure carbon monoxide after lengthy test durations at 1123 K (850°C) has not been observed for any of the three steels. However, the niobium-stabilized steel exhibited a clearly higher mass increase in 100 % CO compared to

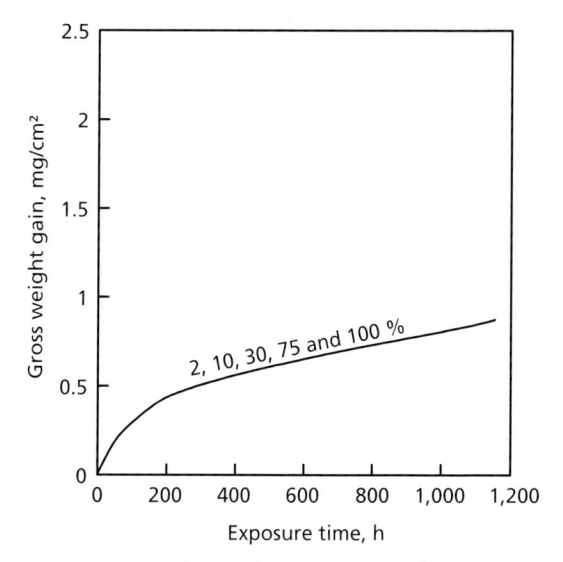

Figure 52: Mass change of 20Cr25Ni/TiN steel as a function of the test duration and the CO concentration at 1123 K (850°C) [98]

test gases with lower CO contents. Practically no influence of the CO content in the gas could be found in the nitrided steel at 1123 K (850°C).

Examinations of the formed cover layers revealed that the mass increase is strongly influenced by the precipitation of carbon. In gases with a higher CO content the niobium-stabilized steel was particularly sensitive to carbon precipitations at all test temperatures, whereas the titanium stabilized steel and in particular the TiN-containing material exhibited a clearly better behavior. The differences in behavior can be explained by the different formation of chromium oxide layers on the surface. Steel 20Cr25Ni/TiN, for instance, forms a chromium oxide-rich cover layer under all test conditions, supported by the fine-grained structure in the surface area and inhibiting the precipitation of carbon. In contrast, more time and a higher temperature were necessary in case of the niobium-stabilized steel 20Cr25Ni/Nb to form a protective cover layer, attributed to its rather coarse-grained structure in the surface area. Regarding its behavior the steel 20Cr25Ni/Ti was ranking between the two other steels. The dependence of carbon precipitation on the CO content of the gas and the test temperature is interpreted as caused by the thermodynamics of the Boudouard reaction.

Due to damage to the fuel element cladding tubes in the nuclear reactors cooled with carbon dioxide gas, the steel with 20% Cr, 25% Ni and 1% Nb used there was intensively examined. The time-related process of oxidation followed a logarithmic time law and, in general, the oxide layer exhibited again a two-layer structure. On the outer side there is an iron-rich spinel layer, whereas there is a Cr_2O_3 layer inside.

As shown in Figure 18 the austenitic 18-8 nickel chromium steels in pure carbon dioxide exhibit a good resistance to oxidation in a temperature range of up to about 803 K (530°C) – to a major extent irrespective of their chemical composition. However, at higher temperatures they are inferior to the steels with higher chromium and nickel contents.

Figure 53 and Figure 54 summarize the test results of the oxidation behavior of austenitic nickel chromium steels in CO-CO_2 mixtures [64].

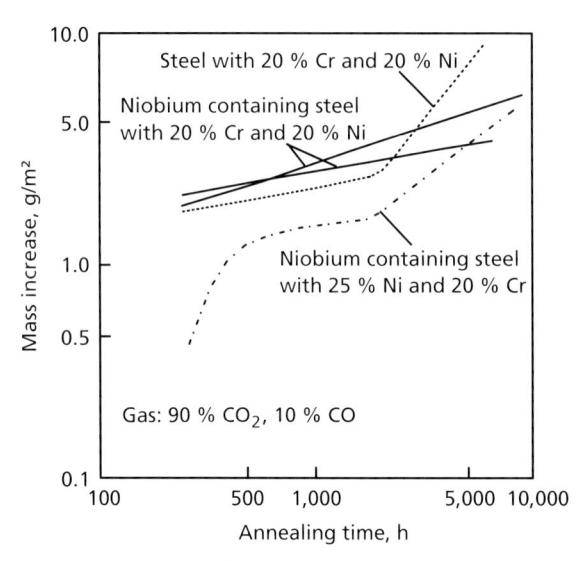

Figure 53: Time-related mass change of various austenitic nickel chromium steels in carbon dioxide with 10 % carbon monoxide at 1023 K (750°C) [64]

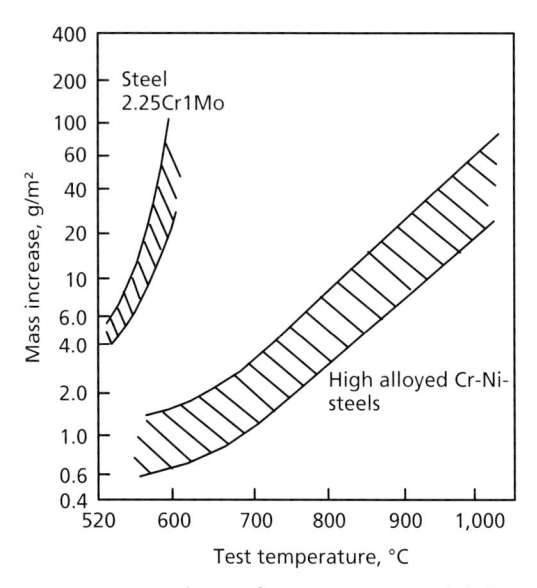

Figure 54: Mass change of various austenitic nickel chromium steels in carbon dioxide with 1 to 10 % carbon monoxide as a function of temperature. Test duration 1000 hours [64]

Comparing Figure 53 and Figure 54 with Figure 18, the addition of carbon monoxide to carbon dioxide reduce the oxidation rates, in particular at lower temperatures. For 18-8 nickel chromium steels the time until the accelerated oxidation starts is clearly shifted to longer times due to the CO content of CO_2.

However, in case of steels with 20 % to 25 % Cr and 20 % to 30 % Ni only slightly lower mass changes were found in CO-CO_2 mixtures compared to pure carbon dioxide after a test duration of 10,000 h as shown in Table 20. [64].

Temperature K (°C)	Gas composition	Mass change g/m²
1023 (750)	100 % CO_2	8 to 25
	90 % CO_2 + 10 % CO	3 to 20
1173 (900)	100 % CO_2	30 to 100
	90 % CO_2 + 10 % CO	20 to 70

Table 20: Mass change of high-alloy austenitic nickel chromium steels in pure carbon dioxide and in carbon dioxide containing 10 % carbon monoxide at different temperatures [64]

In contrast to the low-alloy steels, austenitic nickel chromium steels show a clear influence of the flow rate of carbon dioxide even at lower rates at temperatures of 823 K (550°C) and 873 K (600°C) as shown in Figure 55 using the steel with the DIN Mat. No. 1.4541 [64].

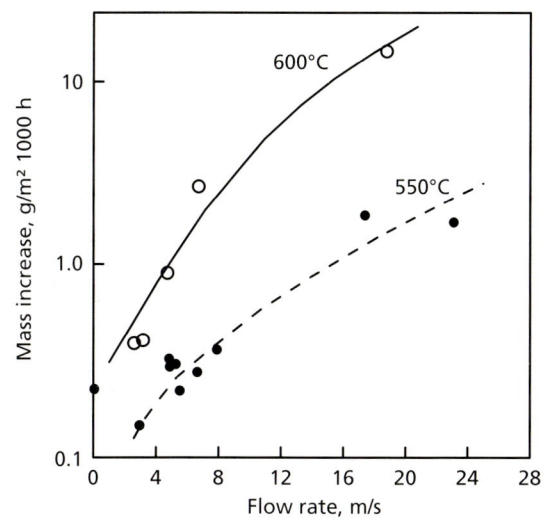

Figure 55: Dependence of the mass change of the steel with the DIN Mat. No. 1.4541 in carbon dioxide on the flow rate of the gas [64]

[100] reports on investigations into the kinetics of high-temperature oxidation and the spalling of the protective coating of an Nb-stabilized 20% Cr-25% Ni steel in carbon dioxide at a pressure of 40 bar and temperatures of 1123, 1143, 1183 and 1223 K (850°C, 870°C, 910°C and 950°C). As the curves in Figure 56 show, the mass increase of the specimens follow a parabolic time law at all test temperatures. The activation energy is indicated to amount to 370 ± 36 kJ/mol.

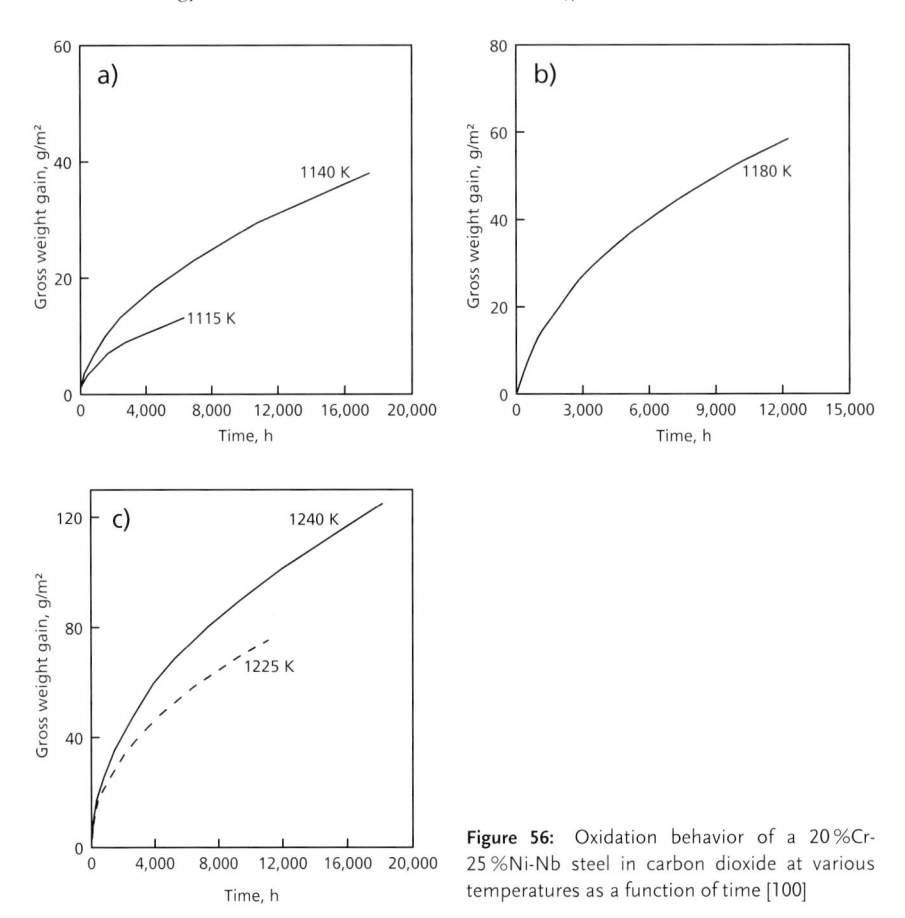

Figure 56: Oxidation behavior of a 20%Cr-25%Ni-Nb steel in carbon dioxide at various temperatures as a function of time [100]

The oxidation rate is determined by the diffusion of the internal chromium oxide layer of a two-layer oxide layer. An intercrystalline oxidation of silicon was observed beneath the oxide layer. The oxide layers spalled off at all temperatures and spalling increased linearly with the mass increase.

The high-temperature oxidation of a niobium-stabilized steel with 25% Ni, 20% Cr in carbon dioxide was examined gravimetrically and also with the help of thin layer activation [101]. During the thin layer activation a defined area of the steel surface was activated by exposure to deuteron radiation to generate radioisotopes of the elements chromium, manganese and cobalt. Measurement of the activity level of the radioisotopes before, during and after oxidation together with the gravimetrical determination of the spalled scale facilitates the assessment of the oxidation processes. The tests were performed under isothermal conditions with a test duration of up to 525 hours in a temperature range between 1023 and 1173 K (750°C and 900°C) and by subsequent cooling in the furnace. The flow rate of the carbon dioxide was 50 cm^3/min. The chemical composition of the examined steel is indicated in Table 21.

C	Si	Mn	Cr	Ni	Nb	Co
0.054	0.62	0.74	19.6	24.7	0.61	0.007

Table 21: Chemical composition of the examined stainless steel [101]

These test results show that at 1173 K (900°C) and 1223 K (950°C) chromium and manganese contribute to the formation of a uniform, stable and protecting oxide layer in the same way. This layer consists of an external spinel layer of an Mn(FeNi-Cr)$_2$O$_3$ composition, a chromium oxide layer (Cr$_2$O$_3$) in the middle and an internal silicon oxide layer. During isothermal oxidation, adhesion of this protective layer is preserved. However, cracking and spalling of the oxide layer occurred after longer oxidation at 1173 K (900°C).

Basically, oxidation processes may also impair the mechanical properties of a material. For instance, the load-bearing section may be reduced if it is uniformly attacked or the internal oxides may act as stress peaks. On the other hand, mechanical loads can also influence the oxidation process by causing defects in the protective oxide layers or causing such layers to spall off.

The investigations described in [102] analyzed the effect of oxidation in carbon dioxide at 1273 K (1000°C) on the grain growth and the long-term resistance of stainless 20Cr-25Ni-Nb steel at 1023 K (750°C), on the one hand, and the effect of elongation under constant load on the oxidation of this steel under the same conditions on the other hand. Examinations were performed with two cold-rolled sheets of different thicknesses with the chemical compositions indicated in Table 22. Prior to the test start, the specimens were solution-annealed in a hydrogen atmosphere at 1203 K (930°C) for 30 minutes and quenched in argon to room temperature.

Thickness	C	Si	Mn	P	S	Cr	Ni	B	N	Nb
0.38 mm	0.045	0.55	0.74	0.005	0.004	19.8	24.4	0.0004	0.005	0.61
0.60 mm	0.026	0.61	0.75	0.005	0.004	20.9	25.0	0.0004	0.005	0.59

Table 22: Chemical composition of the examined sheet specimens [102]

The specimens were oxidized in carbon dioxide at 1273 K (1000°C) and with an ambient pressure for an exposure duration up to 2000 hours. The creep tests were performed with the oxidized specimens and, for comparison, with specimens subjected to heat treatment in air at 1023 K (750°C) by applying a constant load of 46 N. Under the condition, that a cross section reduction did not occur during the oxidation, the initial load of the specimens was 69 MN/m². In another test series the specimens were oxidized under tensile stress in carbon dioxide at 1273 K (1000°C) for 100 hours. The initial load varied from 0 to 25.5 N according to the stresses of 0 to 11.0 MN/m². Table 23 contains the loads of the specimens and details regarding the oxide layers formed.

Stress N	Stress MN/m²	Elongation %	Protective coating thickness μm	Depth of attack μm	Surface without protective coating %
0	0	0	15.5	30	0
0	0	0.5	15.5	43	0
10.5	4.1	3.9	16.6	55	27
12.1	4.9	5.1	14.2	54	50
8.9	3.8	5.2	15.5	55	0
14.5	5.6	9.4	16.6	69	0
22.8	9.9	13.5	16.6	66	60
20.8	8.1	19.7	14.2	70	0
19.2	7.5	19.8	14.2	70	0
24.2	10.5	27.5	14.2	66	8
25.5	11.0	42.5	14.7	75	19
25.5	11.0	51.2	14.7	74	14

Table 23: Load, initial stress and elongation, thickness of the protective coating, depth of attack and share of the surface without protective coating after 100 h oxidation of the specimens in carbon dioxide at 1273 K (1000°C) [102]

As results of the creep tests Figure 57 indicates the creep ductility and the time to fracture of the specimens.

Following the exposure to argon, the ductility (upper section of Figure 57) compared to the initial value of the reheated specimens up to 750 h slightly decreases and remains almost constant thereafter. The ductility of the specimens exposed to carbon dioxide, in contrast, clearly and continuously decreases as the test duration decreases. The lower section in Figure 57 shows that up to about 500 h the lifetime of the specimens stored in argon is slightly higher than the values of the specimens as delivered, the values however slightly decreasing thereafter as the test duration increases. The specimens oxidized in carbon dioxide exhibited identical lifetimes compared to non-oxidized specimens up to 500 h. However, if the test duration is longer, a considerable drop of the lifetime can be observed.

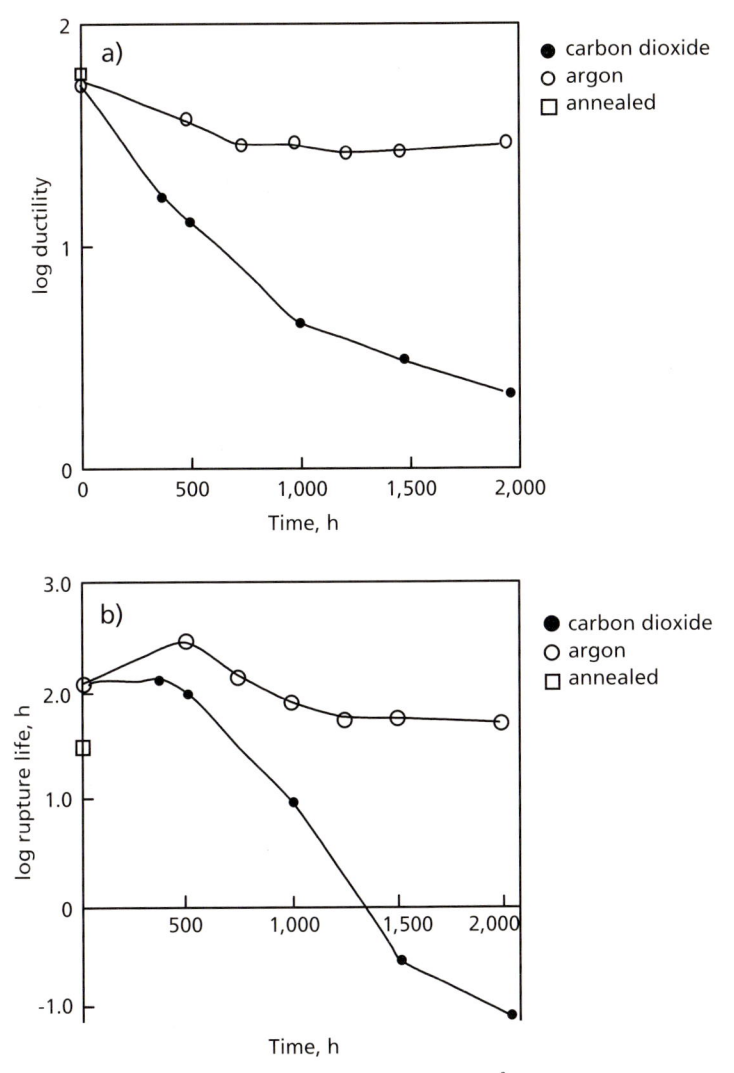

Figure 57: Results of the creep tests (load: 103 MN/m^2) at 1023 K (750°C) in air performed on specimens following an exposure in carbon dioxide (●) or in argon (○) at 1273 K (1000°C) [102]

If exposed to argon at 1273 K (1000°C) for 2000 h, the grain size in the microstructure of the steel clearly grows, but obviously remains without any significant influence on the creep behavior. If exposed to carbon dioxide under the same conditions, the grain size does not grow as is attributable to the precipitation of carbides as a result of the oxidation reaction blocking the grain boundaries and thus preventing growth. But these carbide precipitations lead to embrittlement and therefore to a clear drop in the creep values.

Since the specimens creep at a test temperature of 1023 K (750°C), the specimen cross section changes during the test duration such that the effective stress increases. Table 24 depicts these changes in the test conditions.

Oxidation period h	Effective specimen cross section mm^2	Effective stress MN/m^2
0	0.67	69
500	0.50	93
1000	0.51	90
1500	0.47	97
2000	0.45	103

Table 24: Effective load of the specimens oxidized in carbon dioxide at 1273 K (1000°C) during the creep test at 1023 K (750°C) [102]

Figure 58 shows the relationship between the effective stress and the time to fracture of the specimens starting from an initial stress of 46 N.

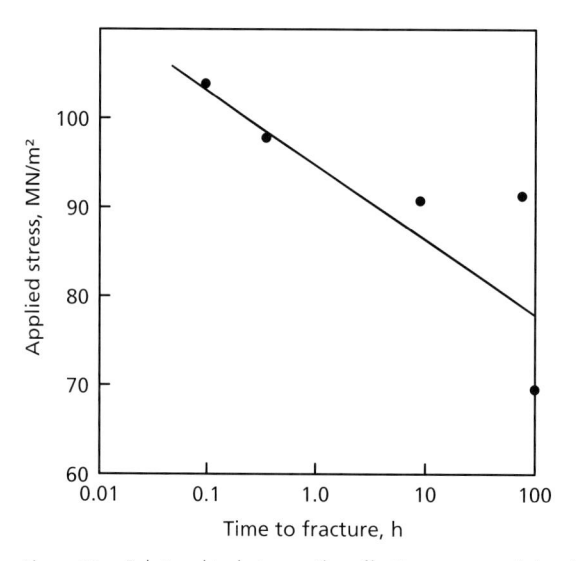

Figure 58: Relationship between the effective stress and the time to fracture of the specimens oxidized in carbon dioxide at 1273 K (1000°C) during the creep test with an initial stress of 46 N [102]

The elongation of specimens resulting from the combined effect of oxidation and loading was determined in the oxidation tests in carbon dioxide at 1273 K (1000°C) for 100 h and concurrent loading of the specimens. As shown in Figure 59 elongation clearly increases as the tensile stress increases.

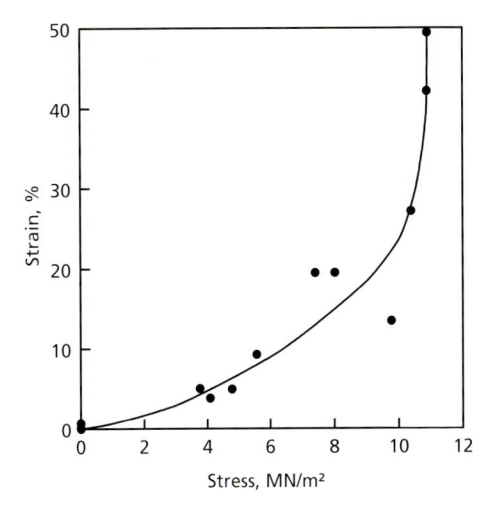

Figure 59: Influence of tensile stress on the elongation of specimens during oxidation in carbon dioxide at 1273 K (1000°C) for 100 h [102]

Although the increasing elongation does not seem to impair the thickness of the uniform, chromium-rich and protecting outer coating, it increases the tendency to imperfections, leads to the formation of non-protecting cover layers and clearly increases the depth of the internal attack.

Chloride and sulfate-containing ash deposits resulting from corrosion by hot combustion gases in power plants can intensify the attack. Since biomass has higher contents of alkaline metals and chlorides, the components in contact with the exhaust gases of biomass-fueled power plants are endangered by ash deposits with high alkaline chloride contents. Therefore, the behavior of the ferritic and austenitic steels used in this area was investigated in simulated combustion gases with and without deposits within the framework of a research project [103].

The composition of the investigated steels is shown in Table 25. The test atmosphere contained 22% H_2O + 5% O_2 + 0 to 25% CO_2 rest N_2. The exposure tests were performed at a temperature of 808 K (535°C) and the test duration was 360 hours.

Steel designation	Cr	Ni	Mo	V	C	Si	Mn	P	S	Nb
2.25Cr1Mo	2.29	0.44	0.96	0.01	0.09	0.23	0.59	0.02	0.02	0.02
X20	10.3	0.72	0.87	0.26	0.18	0.23	0.62	0.02	0.01	
X10	8.7	0.26	0.97	0.23	0.10	0.38	0.48	0.01	0.01	0.07
AC66	27.3	32.2			0.06	0.21	0.64	0.01	0.01	0.78
TP347H	17.6	10.7			0.05	0.29	1.84	0.03	0.01	0.56
Esshete® 1250	15.0	9.65	0.94	0.22	0.08	0.58	6.25	0.02	0.01	0.86
Sanicro® 28	26.7	30.6	3.32		0.02	0.42	1.73	0.02	0.01	

Table 25: Chemical composition of the investigated steels [103]

The results are summarized in Figure 60, Figure 61 and Figure 62. Figure 60 and Figure 62 show that ash deposits clearly increase the corrosive attack both in a CO_2-free and in a CO_2-containing atmosphere. Increasing CO_2 contents in the gas atmosphere intensify the corrosive attack irrespective of ash deposits. The behavior of the austenitic steels was clearly better than that of the ferritic steels in all tests.

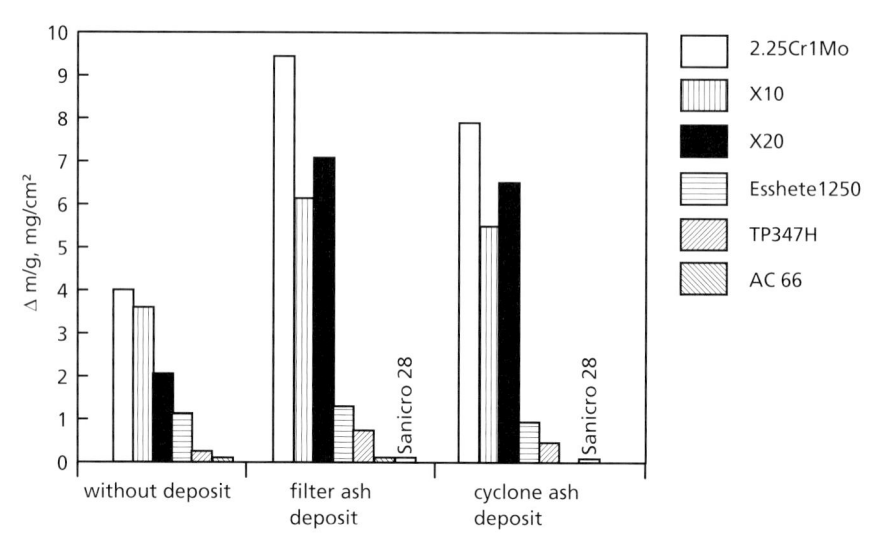

Figure 60: Mass change of the specimens in an atmosphere consisting of 22 % H_2O + 5 % O_2 + N_2 with and without ash deposits after 360 h [103]

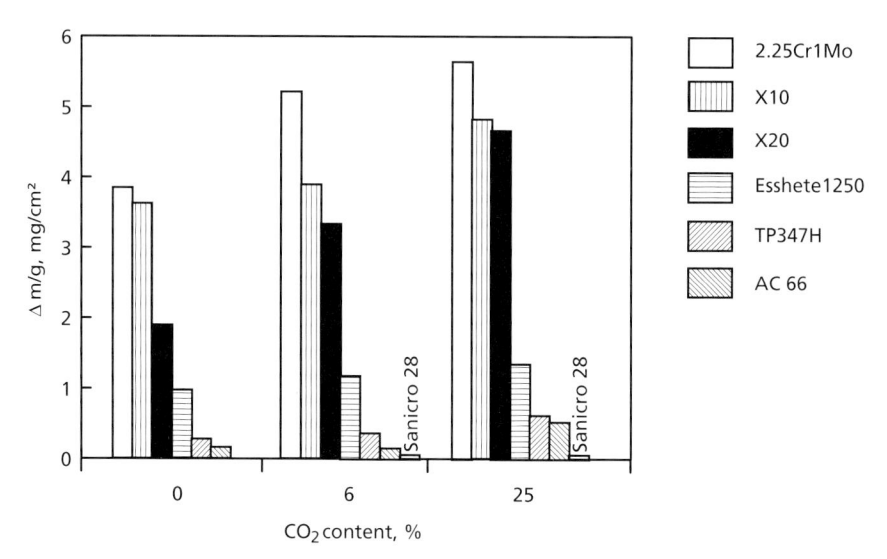

Figure 61: Mass change of the specimens in an atmosphere consisting of 22 % H_2O + 5 % O_2 + x % CO_2 + N_2 without ash deposits after 360 h [103]

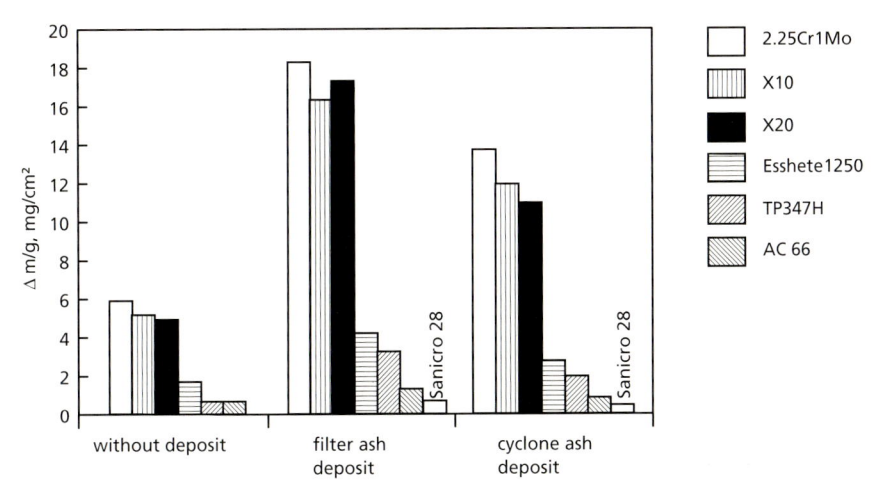

Figure 62: Mass change of the specimens in an atmosphere consisting of 22 % H_2O + 5 % O_2 + 25 % CO_2 + N_2 with and without ash deposits after 360 h [103]

Behavior in aqueous carbon dioxide-containing media

The austenitic stainless Cr-Ni steels exhibit a generally good resistance in aqueous carbon dioxide containing solutions with corrosion rates below 0.01 mm/a (0.40 mpy) [104, 105].

Results of corrosion tests with Cr-Ni steels of different alloy contents in carbonic aqueous solution with a partial pressure of CO_2 of 32.4 bar at 295 K (22°C) are indicated in Table 26 [106].

Steel	Cr %	Ni %	Si %	Corrosion mm/a (mpy)
1	16.8	7.9		0.0025 (0.10)
2	18.2	8.6	2.2	0.0018 (0.07)
3	17.0	23.7	2.9	0.0015 (0.06)

Table 26: Results of corrosion tests with Cr-Ni steels in carbon dioxide-containing solution [106]

Thus, the materials are resistant under the indicated corrosion conditions.

A 21 Austenitic CrNiMo(N) steels
A 22 Austenitic CrNiMoCu(N) steels

Also the molybdenum-containing austenitic CrNi steels of grade AISI 316 or the DIN Mat. No. 1.4404 with about 18 % Cr, 12 % Ni and 2.5 % Mo form cover layers with a multilayer structure in carbon dioxide gas or in gas mixtures of CO_2 and CO at higher temperatures, consisting of Cr_2O_3, Fe_3O_4 and Fe-Cr-Ni spinels depending on the temperature and the gas composition [107].

Occasionally, damage caused by cracks is found in these steels in the temperature range of 773 to 973 K (500°C to 700°C), where oxidation processes play a role. To simulate this crack formation in the lab and examine the influence of the material as well as mechanical and chemical conditions, creep tests and slow strain rate tensile tests were performed with molybdenum-containing and molybdenum-free CrNi steels in different gas atmospheres at 883 K (610°C) [108].

The chemical composition of the investigated steels is shown in Table 27. Four commercial steels (304, 304 H and 316 L, 316 H) with different contents of molybdenum and carbon were used.

Steel	C	Si	Mn	Cr	Ni	Mo
AISI 304	0.04	0.38	1.20	17.8	8.1	0.17
AISI 304 H	0.05	0.36	1.43	18.3	8.6	–
AISI 304 H	0.077	0.30	0.98	18.6	8.5	0.30
AISI 316 L	0.024	–	–	17.1	13.3	2.65
AISI 316 H	0.053	0.47	1.74	16.8	11.0	2.30
AISI 308 H	0.11	0.80	6.60	17.9	8.3	0.10
AISI 309 L	0.02	0.6	1.30	23.5	12.5	0.04
A508C13	0.16	0.2	1.30	0.2	0.7	0.5
X	0.06	0.4	1.3	16.0	9.0	0.15

X = mixed layer of AISI 309 L and A508C13

Table 27: Chemical composition of the investigated steels [108]

The steel 308 H was melted with an elevated manganese content to reach a stable austenitic condition.

To investigate the effect of a grain boundary sensitization by precipitations of chromium carbide ($Cr_{23}C_6$), 304 H steel was subjected to various heat treatment processes to vary the grain size and precipitations (Table 28). As the check for intercrystalline corrosion according to ASTM A 262 Practice B in boiling copper sulfate/sulfuric acid shows, all specimens were sensitized.

In addition, two welding materials were tested, i.e. 308 H steel on the one hand and a hard surfacing layer on the other hand, with 309 L steel as the base material and a cover layer of the unalloyed material A508C13, the contamination of the stainless steel 309 with A508C13 being adjusted to about 30 %.

The tests were performed in vacuum (2×10^{-3} Pa), in air and in a gas mixture of air with 4 % carbon dioxide and 5.8 % or 8 % water vapor.

Designation	Heat treatment	Grain size (ASTM)
G1	state as delivered sensitized at 973 K (700°C) for 30 min	5
G2	state as delivered annealed at 1343 K (1070°C) for 1 hour 5 % cold formed sensitized at 973 K (700°C) for 30 min	¾ to 2
G3	state as delivered sensitized at 883 K (610°C) for 15 h	5

Table 28: Heat treatment and grain sizes of steel 304 H [108]

Creep tests were performed with tensite test specimens of 6 mm in diameter consisting of 304 H steel on lever arm machines and slow strain rate tensile tests were performed with test specimens of 5 mm in diameter at strain rates of 3×10^{-8}, 10^{-7} and 10^{-6}/s for all steels.

The results of the creep tests are summarized in Table 29. Cracks did not occur in air during a test period of 10 h. The extension can be used to assess the creep strength. As expected, the material with the lower grain size is more sensitive than the coarse-grained material.

The microscopic test and the EDX analysis revealed that the cracks formed in the gas mixture during the test are covered by a black oxide layer with a thickness of around 5 µm mainly consisting of iron oxide, probably of magnetite Fe_3O_4. This oxide composition is unusual in a stainless austenitic steel where usually chromium-rich oxides are found.

Designation	Stress MPa	Test medium	Extension %	Crack length µm
G1	200	air	6.4	0
G2	200	air	0.9	0
G1	250	air	15.5	0
G2	250	air	5.4	0
G1	300	air	56	0
G2	300	air	46	0
G1	250	air/CO_2/H_2O	17.3	50
G3	250	air/CO_2/H_2O	23.0	70

Air with 4% carbon dioxide and 5.8% water vapor

Table 29: Results of creep tests [108]

The results of the slow strain rate tensile tests are summarized in Table 30. All cracks in the base materials were merely intercrystalline and interdendritic in the weld specimens. The sensitivity to cracking increases if the strain rate decreases from $10^{-6}/s$ to $10^{-7}/s$.

Steel	Air + 4% CO_2 + 8% H_2O			Air	Vacuum
	Strain rate				
	$10^{-6}/s$	$10^{-7}/s$	$3 \times 10^{-8}/s$	$10^{-7}/s$	$10^{-7}/s$
316 L – 0.024 %C	10	50			
316 H – 0.053 %C		20	50		
304 H – 0.040 %C	10	50	50		
304 H – 0.077 %C		300			
308 H – 0.11 %C	100	750		50	< 10
Hard surfacing layer 1		5000			< 10
Hard surfacing layer 2		5000			

Table 30: Results of slow strain rate tensile tests: Length of the main crack in μm [108]

The higher sensitivity of the hard surfacing layer is attributed to the higher carbon content and the dendritic structure with a higher grain size. The cracks exclusively occurred in the high-alloy part of the hard surfacing layer.

A 24 Magnesium and magnesium alloys

Magnesium and magnesium alloys turned out to be fairly resistant to corrosion in the atmosphere since the resulting corrosion products of magnesium hydroxides and carbonates exert a protecting effect. For the rural environment corrosion rates of about 0.003 mm/a (0.12 mpy) are indicated, for an industrial atmosphere 0.004 mm/a (0.16 mpy) and for a maritime environment 0.006 mm/a (0.24 mpy) [109]. Lab tests focused on the question what the role of carbon dioxide is in the corrosion of magnesium alloys in an atmosphere contaminated with NaCl. Specimens of the magnesium materials with a chemical composition as indicated in Table 31 were exposed to a simulated maritime atmosphere for four weeks.

Material	Al %	Zn %	Mn %	Si %	Fe %	Cu %	Ni %
AM20	2.1	0.04	0.40	0.01	0.0017	0.0016	0.0005
AM60	6.0	0.01	0.25	0.01	0.0016	0.0010	0.0007
AZ91	8.9	0.74	0.21	0.008	0.0022	0.0007	0.0004

Remainder: Magnesium

Table 31: Chemical composition of the investigated magnesium alloys [109]

The test apparatus was set to a relative humidity of 95 % and a temperature of 295 K (22°C). Either the test atmosphere contained less than 1 ppm CO_2 or 350 ppm of pure CO_2 were fed. Prior to the exposure 70 µg/cm² NaCl were applied to the specimen surface by spraying on a solution of common salt. The NaCl layer applied and dried before exposure was gravimetrically determined. The mass change of the specimens was determined once per week without previously drying the corrosion products. The results of the moist mass increases are plotted in Figure 63 against the exposure duration.

As expected, moist air in the presence of chlorides exerts a very aggressive influence on magnesium materials. The very high mass increase at the beginning of the exposure is probably attributable to water absorption of the dried NaCl on the surface. In correspondence with the results reported in the literature from salt spray tests, the following sequence applies to the mass increase: AZ91 – AM60 – AM20, i.e. it rises as the content of the aluminium and zinc alloying elements decreases in the examined magnesium materials.

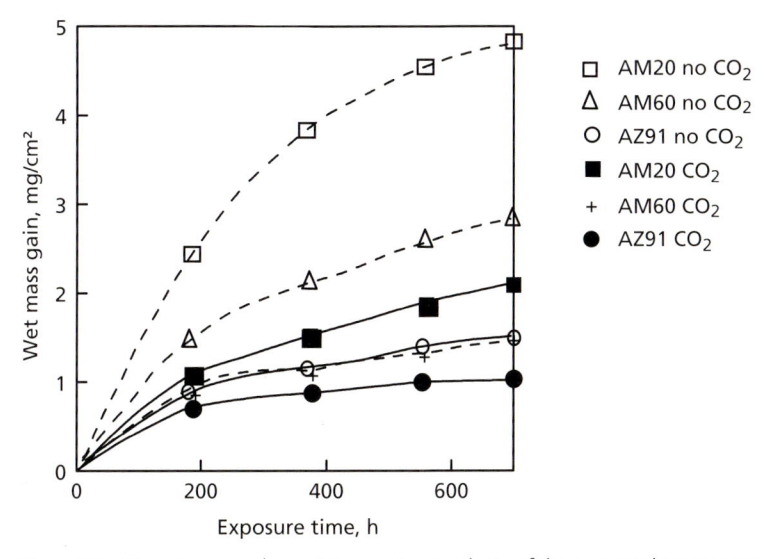

Figure 63: Mass increases by moist corrosion products of the examined magnesium alloys as a function of the test duration [109]

Following the end of the test, the specimens were dried and the mass change was determined again. Then, the corrosion products were removed and the material consumption of the specimens was determined. The results are summarized in Table 32.

The presence of carbon dioxide in the test atmosphere exerts a clear influence on the corrosion behavior. It is not only the mass increase, which is clearly lower in the presence of carbon dioxide, as shown in Figure 63, but as the results in Table 32 show also the general corrosion rate of the three alloys was increased by the factor of about 4 in the absence of carbon dioxide compared to the CO_2 containing atmosphere.

Material	CO$_2$ content ppm	Mass increase dry mg/cm^2	Material consumption mg/cm^2	Corrosion rate mm/a (mpy)
AM20	350	1.9	0.70	0.052 (2.04)
AM60	350	1.3	0.49	0.037 (1.46)
AZ91	350	0.70	0.23	0.015 (0.59)
AM20	0	4.1	3.2	0.24 (9.45)
AM60	0	2.5	1.6	0.12 (4.72)
AZ91	0	1.0	0.64	0.045 (1.77)

* Material consumption after removal of the corrosion products

Table 32: Results of exposure tests [109]

In the CO$_2$-free atmosphere the corrosion products mainly consist of Mg(OH)$_2$ and lower contents of Mg$_{10}$(OH)$_{18}$Cl$_2$ · 5 H$_2$O, whereas in the presence of carbon dioxide mainly the Mg$_5$(CO$_3$)$_4$(OH)$_2$ · 5 H$_2$O compound is formed. In addition also Mg(OH)$_2$ was found on the specimens of the AM20 and AM60 alloys.

The surface of the specimens exposed to the CO$_2$-containing atmosphere was covered by a thick layer of the mentioned corrosion products and simultaneously exhibited a uniform surface attack. Areas with flat pitting corrosion sites were only found on the specimens of both alloys with the lower aluminium content. On the other hand, severe pitting corrosion occurred on the entire surface of the specimens exposed to the CO$_2$-free atmosphere. Obviously, the presence of carbonate layers formed in the presence of carbon dioxide exerts a protecting effect with regard to the atmospheric corrosion of magnesium and its alloys.

However, the formation of such protecting surface layers cannot be expected in aqueous carbon dioxide solutions and the free carbonic acid formed may lead to a corrosive attack on magnesium and magnesium alloys.

A 25 Molybdenum and molybdenum alloys

Molybdenum and molybdenum alloys are not attacked in carbon dioxide-containing aqueous solutions.

A 26 Nickel

A minor consumption corresponding to a corrosion rate of 0.002 mm/a (0.08 mpy) was measured in carbon dioxide-containing waters with a partial pressure of CO$_2$ of 13.8 bar and a temperature of 328 K (55°C) after an exposure duration of 7 days. After longer test durations of 28 and 70 days the values were even lower, implying the formation of a passivating surface layer [106].

Also according to older examinations nickel with a corrosion rate of 0.05 mm/a (1.97 mpy) is considered resistant in pure water at a carbon dioxide pressure of

2.5 bar and a temperature of 343 K (70°C). Specifications of manufacturers also confirm resistance in carbon dioxid-containing waters even at temperatures up to 723 K (450°C) and in the concurrent presence of sulfur dioxide.

A 27 Nickel-chromium alloys
A 28 Nickel-chromium-iron alloys (without Mo)

In carburizing gases containing carbon dioxide, carbon monoxide or hydrocarbons, the behavior of Ni-Cr or Ni-Cr-Fe alloys differs. In CO_2, for instance, not only damage to the material by carburization but also by oxidation must be expected. Concurrent carburization and oxidation impede the formation of a protecting oxide layer such that Ni-Cr and Ni-Cr-Fe alloys should only be used in this atmosphere up to 1273 K (1000°C).

In carbon monoxide atmosphere the so-called green rot of Ni-Cr alloys can be observed in a temperature range from 1088 to 1253 K (815°C to 980°C) caused by the inner selective oxidation of chromium. The partial oxygen pressure of the atmosphere is so low that, in particular, nickel oxide playing a special role in the initial oxidation phase cannot be formed.

Therefore, Ni-Cr-Fe alloys are used in such media since the iron oxide is resistant also under the low partial oxygen pressures occurring here and hence can contribute to the protective effect of the material [110]. Alloying additions of silicon lead to a clear reduction of the corrosive attack in carbon monoxide atmospheres.

Generally, Ni-Cr-Fe alloys with 14 % to 16 % chromium and about 6 % iron permit operating temperatures up to 1273 K (1000°C) without any significant internal carburization or oxidation occurring. Chromium contents of 20 % up to 22 % lead to a clear increase of the corrosive attack in the temperature range of chromium carbide precipitation (about 1073 K (800°C)). Silicon or aluminium-alloyed Ni-Cr-Fe materials can be used even at temperatures above 1273 K (1000°C). Both alloying elements form protecting oxide layers reducing the carbon diffusion.

A 29 Nickel-chromium-molybdenum alloys

The nickel alloy Alloy 825 (Incoloy® 825), i.e. a titanium-stabilized alloy, does not only exhibit a good corrosion resistance both under oxidizing and reducing conditions, but also has a good resistance to hot gases, e.g. combustion gases with sulfur dioxide and steam. The work described in [111] examined the behavior of 1 mm thick sheet specimens of Alloy 825 in water vapor, synthetic air (20/21 % oxygen in nitrogen) and pure carbon dioxide as well as in a carbon dioxide atmosphere diluted with argon by the factor 100. The composition of the specimens is specified in Table 33. The tests were performed at 1073, 1273 and 1473 K (800°C, 1000°C and 1200°C). The test duration was between 15 minutes and 24 hours.

C	Si	Cu	Fe	Mn	Cr	Ti	Al	Co	Mo	Ni
0.01	0.24	2.00	31.10	0.11	22.00	0.98	0.12	0.67	3.29	39.58

Table 33: Chemical composition of the examined Alloy 825 [111]

Figure 64 summarizes the results of the mass increase in double logarithmic representation.

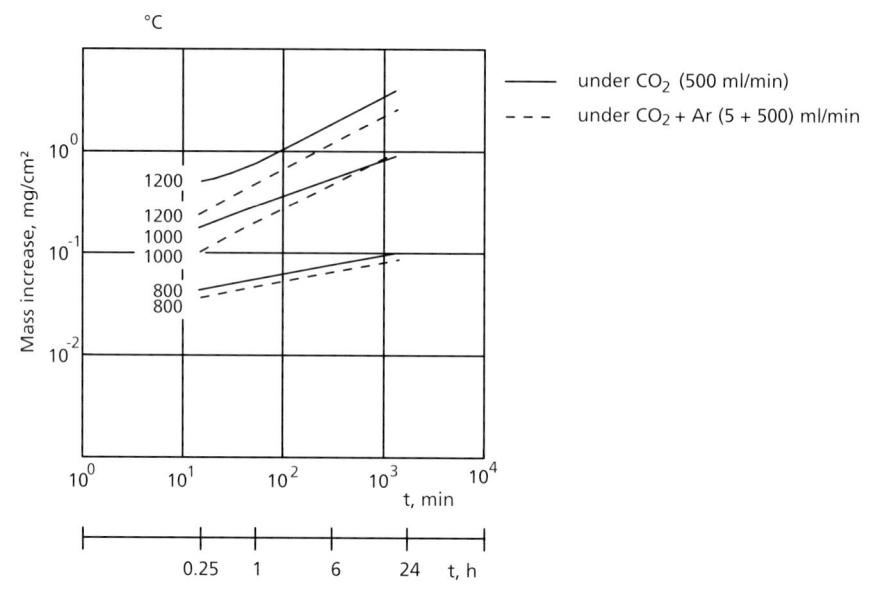

Figure 64: Mass increase of the Alloy 825 specimens in pure carbon dioxide and carbon dioxide diluted with argon (1:100) as a function of temperature and exposure duration [111]

Dilution with inert gas noticeably reduces the oxidation of the nickel alloy compared with the reaction in pure carbon dioxide. In pure carbon dioxide the reaction rate at 1073 K (800°C) is lower than corresponding to a quadratic time dependence, follows a quadratic time law at 373 K (100°C), whereas the reaction kinetics are determined by a parabolic time law at 1473 K (1200°C).

As the parallel curve path shows, the reaction kinetics in diluted carbon dioxide at 1073 K (800°C) corresponds to that in pure carbon dioxide. If the temperature is increased, the time dependence of oxidation is shifted into the direction of a parabolic curve path.

At 1073 K and 1273 K (800°C and 1000°C) firmly adhering cover layers are formed, whereas a part of the layers spalls off at 1473 K (1200°C). Spalling is higher after short test times than after longer exposure durations since the increasing inner oxidation and the formation of voids result in the formation of thicker and more uniform layers. The improvement of adhesion and thermal shock resistance of the layer observed with increasing exposure duration prevents or impedes spalling. It is assumed that the voids provide enough space to absorb the increase in volume of the growing layer.

In aqueous carbon dioxide-containing solutions the passivatable NiCr, NiCrFe and NiCrMo alloys are practically resistant. Stress corrosion cracking of these materials is not even found in CO_2/H_2S-containing solutions [112].

A 33 Lead and lead alloys

Lead and lead alloys are attacked by carbonated waters if they contain oxygen and if a sufficient protective layer is not formed.

The corrosive attack of technical process waters containing carbon dioxide by lead can be reduced to sufficiently low values by means of an adequate water treatment.

A 34 Platinum and platinum alloys
A 35 Platinum metals (Ir, Os, Pd, Rh, Ru) and their alloys

Platinum and platinum metals as well as their alloys are resistant to aqueous carbon dioxide-containing solutions under all conditions.

A 36 Tin and tin alloys

Tin is hardly attacked by aqueous carbonic acid solutions up to temperatures of 373 K (100°C). Even at higher carbon dioxide contents the corrosion rates are below 0.01 mm/a (0.39 mpy).

A 37 Tantalum, niobium and their alloys
A 38 Titanium and titanium alloys

The highly corrosion resistant materials of groups A 37 and A 38 are resistant to aqueous carbon dioxide-containing solutions under all conditions.

A 39 Zinc, cadmium and their alloys

In the corrosion products which are formed on zinc surfaces in the atmosphere and which are deemed to have protecting properties, hydroxy carbonates, mostly hydrozincite $Zn(CO_3)_2(OH)_6$ is detected. This shows that carbon dioxide plays an important role during the corrosion of zinc in a free atmosphere. [113] examined the influence of carbon dioxide on the corrosion in humid air and the corrosion products formed and described their protective effect. For tests performed in a lab apparatus with a controlled gas flow of 1000 ml/min specimens of electrolytic zinc were chosen with a purity of > 99.9 %. The test temperature was 295 K (22°C) and the relative humidity in the apparatus was adjusted to 95 % by mixing of dry air with water vapor.

With contents of < 1 ppm, 350 ppm, 1000 ppm and 40 000 ppm, four different carbon dioxide concentrations were adjusted. The exposure duration was four weeks and once per week the mass change of the specimens was determined. Figure 65 shows the mass increases as a function of time.

As expected, the corrosion rate increases as the carbon dioxide content in the atmosphere increases. The corrosive effect of carbon dioxide is attributed to an acidification of the electrolytic film on the surface. In addition to hydrozincite, the com-

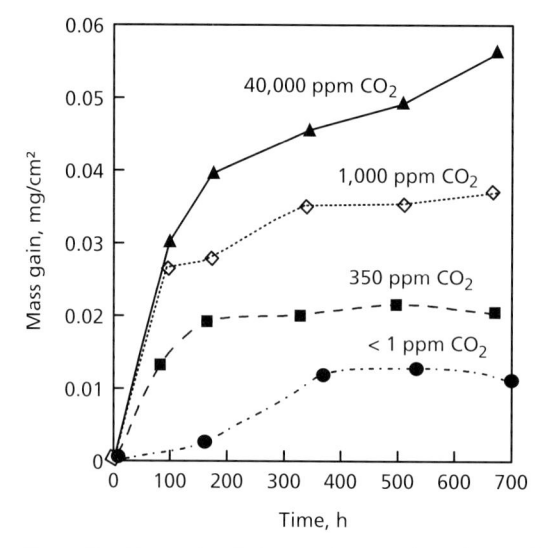

Figure 65: Mass gain of zinc specimens in humid air with 95 % relative humidity and different contents of carbon dioxide [113]

pound $Zn_4CO_3(OH)_6 \cdot H_2O$ was found as a main component of the corrosion products at high carbon dioxide concentrations, whereas at lower carbon dioxide contents ZnO was formed.

Under the condition that only ZnO is formed in a CO_2-free atmosphere and only $Zn_4CO_3(OH)_6 \cdot H_2O$ is formed in a CO_2-containing atmosphere, corrosion rates are calculated from the mass increases indicated in Table 34.

CO_2 content Vol.-ppm	Mass increase µg/cm²	Calculated material consumption µg/cm²	Calculated corrosion rate µm/a (mpy)
0	5	20	0.3 (0.012)
350	15	22	0.4 (0.016)
1 000	28	41	0.7 (0.027)
40 000	47	68	1.2 (0.047)

Table 34: Calculated corrosion rate of pure zinc in air with different carbon dioxide contents with a relative humidity of 95 % and a temperature of 295 K (22 °C) after a test duration of four weeks [113]

Both the cover layers formed in a CO_2-free atmosphere and the cover layers formed in the presence of carbon dioxide, both consisting of zinc oxide or zinc hydroxy carbonates, have no protecting effect if the atmosphere is contaminated with SO_2 as revealed by subsequent tests during which sulfur dioxide was subsequently added to the test atmosphere. This is shown in Figure 66 for the examples of specimens exposed to CO_2-free air (section (a) of the figure), on the one hand, and to air with 1000 ppm CO_2 (section (b) of the figure), on the one hand, for 200 h, followed by an addition of 225 ppb SO_2 to their test atmosphere.

Irrespective of the corrosion product in the beginning, in the presence of SO_2 the compound $Zn_4CO_3(OH)_6 \cdot H_2O$ is formed as the main component of the cover layer.

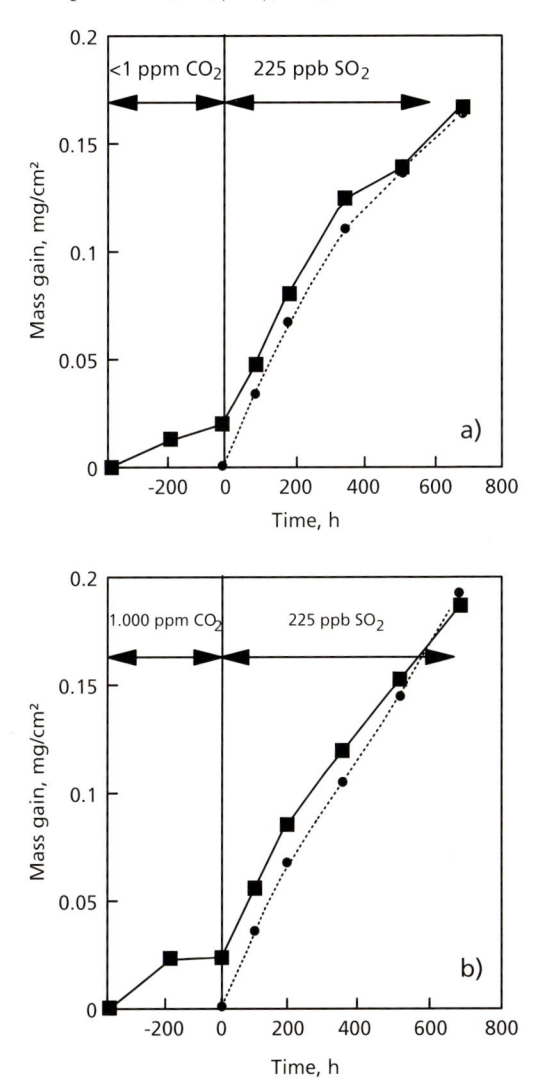

Figure 66: Mass gain of zinc specimens as a function of the test duration following a pre-exposure to CO_2-free (a) and CO_2-containing air (b) and addition of 225 ppb SO_2 to the test atmosphere [113]

A 40 Zirconium and zirconium alloys

The highly corrosion resistant materials of groups A 40 are resistant to aqueous carbon dioxide-containing solutions under all conditions.

B
Nonmetallic inorganic materials

B 3 Carbon and graphite

Material (Trade name[1])	Medium	Concentration	Temperature K (°C)	Resistance	References
a) Manufacture of chemical equipment					
Graphite, phenolic resin impregnation (Diabon® N, NS1; NS2)	aqueous CO_2 solution	≤ saturated ≤ saturated	≤ 453 (≤ 180) ≤ 473 (≤ 200)	+ +	[114] [114]
	CO_2 gas, moist	any any	≤ 453 (≤ 180) ≤ 473 (≤ 200)	+ +	[114] [114]
	CO_2 gas, dry	any any	≤ 453 (≤ 180) ≤ 473 (≤ 200)	+ +	[114, 115] [114]
Graphite, polytetrafluoroethylene (PTFE) impregnation (Diabon® CT)	aqueous CO_2 solution	≤ saturated	≤ 473 (≤ 200)	+	[114]
	CO_2 gas, moist	any	≤ 473 (≤ 200)	+	[114]
	CO_2 gas, dry	any	≤ 473 (≤ 200)	+	[114]
Graphite, polyvinylidene fluoride (PVDF) impregnation (Diabon® F 100)	aqueous CO_2 solution	≤ saturated	≤ 413 (≤ 140)	+	[114]
	CO_2 gas, moist	any	≤ 413 (≤ 140)	+	[114]
	CO_2 gas, dry	any	≤ 413 (≤ 140)	+	[114]

+ = resistant − = not resistant
RT = room temperature
1) Trade name nonbinding

Table 35: Resistance of graphite materials to aqueous carbon dioxide solutions, liquid and gaseous carbon dioxide

Table 35: Continued

Material (Trade name[1])	Medium	Concentration	Temperature K (°C)	Resistance	References
b) Storage technology					
Carbon					
– without impregnation	CO$_2$ gas	100 %	RT	+	[116]
	liquid CO$_2$	100 %	RT	+	[116]
– synthetic resin impregnation	CO$_2$ gas	100 %	RT	+	[116]
	liquid CO$_2$	100 %	RT	+	[116]
– (PTFE) impregnation	CO$_2$ gas	100 %	RT	+	[116]
	liquid CO$_2$	100 %	RT	+	[116]
– metal impregnation	CO$_2$ gas	100 %	RT	+	[116]
	liquid CO$_2$	100 %	RT	+	[116]
c) High-temperature applications					
Graphite	CO$_2$ gas	100 %	898 (625)	+	[117]
			> 1173 (> 900)	–	[117, 118]
Carbon fiber reinforced graphite (Sigrabond®)	CO$_2$ gas	100 %	> 1173 (> 900)	–	[118]

+ = resistant – = not resistant
RT = room temperature
1) Trade name nonbinding

Table 35: Resistance of graphite materials to aqueous carbon dioxide solutions, liquid and gaseous carbon dioxide

B 4 Binders for building materials (e.g. concrete, mortar)

Carbonatization, i.e. the reaction of carbon dioxide from the atmosphere with hydrated Portland cement, the common binding agent of concrete, causes undesired changes in concrete. The basic reaction is the conversion of calcium hydroxide with carbon dioxide forming calcium carbonate and water:

$$CO_2 + Ca(OH)_2 \rightarrow CaCO_3 + H_2O$$

Carbonatization causes the concrete to shrink, which may lead to the formation of cracks, in particular if shrinkage is intensified by evaporation of water (drying) of the concrete. In steel-reinforced concrete carbonatization leads to a reduction of the pH value and hence to a loss of the passivating effect of the concrete such that the corrosion of the steel is enormously supported. The rate of carbonatization strongly depends on humidity: dry concrete or concrete at 100 % relative air humidity with its pores filled with water practically does not react with carbon dioxide at all; the highest reaction rate is observed at a relative air humidity of about 50 %. However, apart from these undesired changes the reaction with carbon dioxide can also be used to reach specific positive property changes, e.g. a harder concrete surface or higher compressive strength [119].

If (running) natural water having absorbed carbon dioxide from the air acts on concrete components, it must be expected that soluble contents of the concrete are washed out. A major soluble content of concrete is calcium hydroxide, which is converted to calcium carbonate according to the above acid-alkali reaction. The amount of washed-out calcium hydroxide or calcium carbonate depends on the temperature, the CO_2 content and the hardness of the natural water; cold water is more aggressive than warm water and soft water causes a stronger attack than hard water because in both cases calcium hydroxide is dissolved more easily [119].

In [120] the influence of the cement content on carbonatization and other transport processes in concrete was investigated. The cement content was varied from 300 kg/m^3 to 450 kg/m^3 and two water/cement ratios (0.4 and 0.5) were used. After curing disk-shaped specimens (diameter: 100 mm, thickness: 50 mm) were stored in air with a relative humidity of 50 % at 303 K (30°C) for 25 weeks and subsequently, after the rounded surfaces had been sealed with a pressure-sensitive adhesive tape, exposed to air containing 5 % CO_2 and having a relative humidity of 50 % for 20 weeks at 303 K (30°C). The carbonatization depth measured at the specimen cross sections stained with phenolphthalein increased if the cement content was increased and, with identical cement content, it was always clearly higher in the specimens with a water/cement ratio of 0.5 compared to the specimens with the lower water/cement ratio.

The influence of the cement content and the water/cement ratio on the carbonatization rate is also described by investigations in [121]. In addition, part of the cement was substituted by ground, granulated blast furnace slag of different particle size, fly ash or silica fume to subsequently investigate the influence of mineral additives on carbonatization. The composition of the concrete specimens is indicated in Table 36. Carbonatization was monitored over a period of one year, using prismatic specimens (100 mm × 100 mm × 140 mm) hardened at 303 K (30°C) in water saturated with calcium carbonate and exposed to air containing 6.5 % CO_2 with a relative humidity 65 % at 303 K (30°C). There was a linear relationship between the carbonate penetration depth X in mm and the square root of the penetration time t in weeks:

$$X = (C \sqrt{t}) + a$$

whereas C is the carbonatization coefficient in mm/√weeks and a is an empirical constant in mm, determined by linear regression from the measured values. The compilation of the carbonatization coefficients in Table 36 reveals that the carbonatization rate of the types of concrete without mineral additives increased when the cement content was reduced and the water/cement ratio was increased, whereas specimen 1 with the highest cement content did not show any carbonatization at all. Noticeable reduction of the carbonatization coefficient and hence the carbonatization rate of the concrete specimen 3 by mineral additives could only be detected if a rather high portion of the concrete was replaced by blast furnace slag types with a particularly small particle size (specimens 8 and 9) or when a small portion of the cement was replaced silica fume (specimen 12).

No.	Cement content kg/m^3	Mineral additive			Water/ cement ratio	Carbonati- zation coefficient mm/$\sqrt{\text{weeks}}$
		Designation	Specific surface cm^2/g	Degree of substitution of the cement Ma%		
1	500	without polymer addition	–	0	0.3	0
2	400	without polymer addition	–	0	0.4	0.782
3	350	without polymer addition	–	0	0.5	2.624
4	300	without polymer addition	–	0	0.6	4.829
5	245	ground, granulated blast furnace slag	4500	30	0.5	2.469
6	175	ground, granulated blast furnace slag	4500	50	0.5	2.932
7	123	ground, granulated blast furnace slag	4500	65	0.5	3.248
8	123	ground, granulated blast furnace slag	6000	65	0.5	2.068
9	123	ground, granulated blast furnace slag	8000	65	0.5	1.678
10	298	fly ash		15	0.5	2.667
11	245	fly ash		30	0.5	3.105
12	333	silica fume	250000	5	0.5	2.325
13	315	silica fume	250000	10	0.5	2.534

Filler content of concrete: about 1810 kg/m^3
Test conditions: Temperature: 303 K (30°C); CO$_2$ content: 6.5 %; Relative humidity: 65 %

Table 36: Carbonatization coefficient of concrete specimens with different composition [121]

The carbonatization of concrete is not only accompanied by a reduction of the pH value and, therefore, the loss of depassivation of the reinforcing steel, but also by changes of the permeability and the pore volume of the concrete as shown by investigations in [122]. Prismatic specimens (50 mm × 50 mm × 300 mm) of two types of concrete, i.e. a high-quality concrete (water/cement ratio: 0.45; Portland cement content: 400 kg/m^3; Compressive strength: 50.0 N/mm^2) and a type of concrete of a lower quality (water/cement ratio: 0.60; Portland cement content: 300 kg/m^3; Compressive strength: 23.0 N/mm^2), were cured either in air or water for 28 d; thereafter, air containing 5% CO_2 was applied to two opposite sides of the specimens for a period of up to 140 d. After different test periods the carbonatization depth was determined on the specimen cross sections by staining with phenolphthalein, whereas permeability was measured by means of a specifically developed vacuum apparatus and the total pore volume was determined in the carbonated area using helium after the water had been removed with isopropanol. As the time of CO_2 penetration increased, the carbonatization depth increased in all specimens, whereas permeability and the total pore volume were reduced. The most marked changes were detected in specimens of a lower quality concrete cured in air, followed by the specimens cured in water. The specimens of the high-quality concrete cured in air or in water showed a much more favorable behavior, i.e. the carbonatization rate, the reduction of permeability and the pore volume were lower, the specimens cured in water achieving a slightly better result compared to the specimens cured in air.

Variable amounts of aqueous polymer dispensions or water-dilutable epoxy resins were added to Portland cement bound mortars and after the setting process the mortar specimens were exposed to carbon dioxide-containing air to study the influence of the polymer additives on the carbonatization rate [123]. The mortar compositions contained calcium carbonate and three quartz sands as fillers (grain size: < 0.5 mm, 0.5–2 mm and 2–4 mm); the water/cement ratio was adjusted from 0.40 to 0.55. Six different polymer dispersions were considered, including dispersion on the basis of polymethylmethacrylate, polybutylacrylate or vinyl acetate copolymers, and 4 different water-dilutable, two-component epoxy resins, 5 to 20% by volume of the concrete being replaced by the polymers. For determining the carbonatization rate air containing 3% carbon dioxide and having a relative humidity of 50 to 70% was applied to prismatic mortar specimens (40 mm × 40 mm × 160 mm) after they had been cured for 12 weeks, at 293 K (20°C) up to a period of 40 weeks. The carbonatization depth was determined on the specimen cross sections by staining with phenolphthalein as a function of the test duration. The mortar specimens with the water/cement ratio of 0.55 exhibited a higher carbonatization depth than the specimens with a lower water/cement ratio. None of the polymers was able to significantly reduce the carbonatization rate or depth. Unexpectedly, several commercial epoxy resin systems yielded clearly poorer results than the comparative cement mortar without polymer additive.

B 5 Acid-resistant building materials and binders (putties)

Pipelines of polymer concrete containing about 15% of an unsaturated polyester resin (UP resin) as a binding agent, are resistant to aqueous carbon dioxide solutions at ambient temperature [124].

B 8 Enamel

For enamel coatings a resistance to carbon dioxide is indicated up to 453 K (180°C) [125].

The dependence of the resistance of a heavy metal free enamel coating (Pfaudler® PharmaGlass PPG), that is specially recommended for pharmaceutical production purposes, on the pH is shown in Figure 67 [126]. For penetrating aqueous carbon dioxide solutions the maximum corrosion rate in the pH range 4 to 6 is 0.1 mm/a (3.94 mpy) up to a temperature of about 433 K (160°C). For another enamel quality a resistance of up to 413 K (140°C) is indicated if aqueous carbon dioxide solutions with a CO_2 content of ≥ 200 mg/l penetrates the material [127].

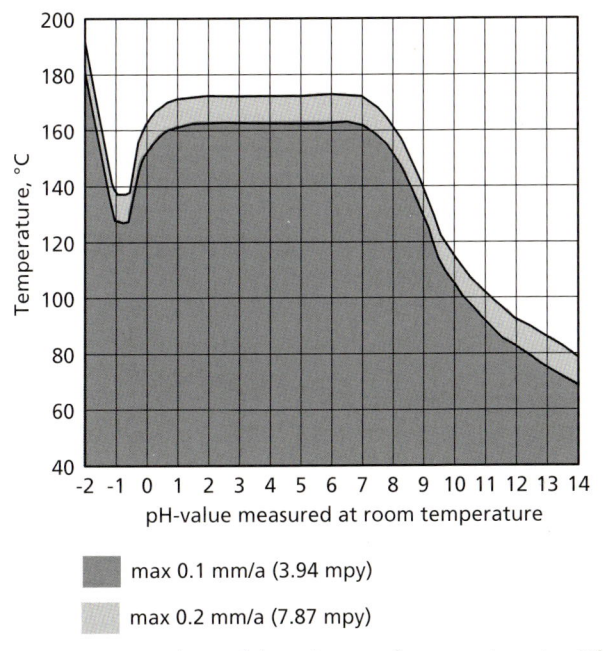

■ max 0.1 mm/a (3.94 mpy)

▨ max 0.2 mm/a (7.87 mpy)

Figure 67: Dependence of the resistance of an enamel coating (Pfaudler PharmaGlass PPG) on the pH value [126]

B 12 Oxide ceramic materials

Ceramic materials	Temperature K (°C)	Resistance	References
Aluminium oxide	1473 (1200)	+	[117]
Magnesium oxide	1473 (1200)	+	[117]
Beryllium oxide	1473 (1200)	+	[117]
Zirconium dioxide, stabilized	1473 (1200)	+	[117]

+ = resistant

Table 37: Resistance of oxide ceramics to carbon dioxide gas

B 13 Metal-ceramic materials (Carbides, Nitrides)

The very good oxidation resistance of silicon carbide (SiC) at high temperatures results from the formation of a dense and smooth SiO_2 layer which hinders or prevents the ongoing oxidation. In gas atmospheres with a lower oxidation potential the formation of an SiO_2 layer is more difficult and volatile SiO compounds are formed leading to an active corrosion and higher material consumption. Using SiC in hot flue gases in practice, the oxidation behavior of these components depend on the content of elements with an oxidizing effect, such as oxygen, carbon dioxide and water vapor.

[128] reports about the oxidation behavior of SiC layers with a thickness of 1 mm in CO-CO_2 atmospheres in a temperature range from 1823 to 1923 K (1550°C to 1650°C). The SiC layers were separated by means of CVD (chemical vapor deposition) on graphite plates. First the specimens were heated in flowing argon up to 1573 K (1300°C) and then adjusted to the desired temperature in flowing CO-CO_2 mixtures, where they were kept for 4 h or 8 h. The ratio of the partial pressures $P_{(CO_2)}/P_{(CO)}$ was varied between 10^{-4} and 10^{-1}. The specimens were cooled down to 1573 K (1300°C) with a cooling rate of 3 K/min in the test atmosphere and subsequently in argon. The mass changes of the specimens were thermogravimetrically determined. The material consumption rates followed a linear time law according to Equation 16 as shown in the example in Figure 68.

Equation 16 $\quad M = k_{(CO/CO_2)} \cdot t$

M = mass loss of the specimens, t = time, $k_{(CO/CO_2)}$ = mass loss rate

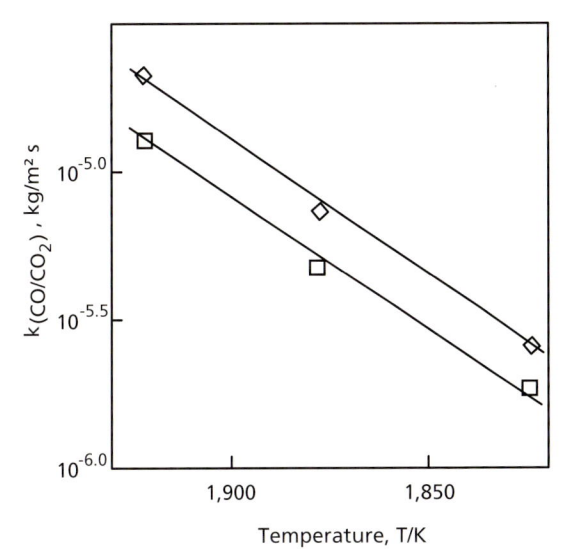

Figure 68: Time dependence of the mass loss rates $k_{(CO/CO_2)}$ of SiC at two different partial pressure ratios $P_{(CO_2)}/P_{(CO)}$ ($\diamond$ $P_{(CO_2)}/P_{(CO)} = 10^{-2.5}$; $\square$ $P_{(CO_2)}/P_{(CO)} = 10^{-2.0}$) [128]

The maximum mass loss rates were observed at a partial pressure ratio $P_{(CO_2)}/P_{(CO)}$ in a range from $10^{-2.5}$ to 10^{-3} depending on the temperature or flow rate of the gas.

[129, 130] provides an overview over the oxidation behavior and the high-temperature corrosion of silicon carbide as pure powders as well as in hot pressed and sintered form. To summarize this literature overview, the attack of CO- and CO_2-containing gases corresponds to an oxidation under reduced partial oxygen pressure with a dissociation of CO_2 to form CO and $1/2\ O_2$. Consequently, the oxidation rates are lower than in oxygen.

The kinetic mechanisms are determined by the permeation of oxygen and the back-diffusion of the formed CO. The activation energy amounts to 126.5 kJ/mol for CO_2 and 204.3 kJ/mol for 2.5 % CO_2.

Temperatures above 1723 K (1450°C) lead to the decomposition of the SiO_2 cover layer. The application limit for SiC in CO_2 containing air is 1623 K (1350°C). During the oxidation in CO volatile SiO is formed at temperatures from 1623 K (1350°C). At temperatures below 1523 K (1250°C) SiC is no longer attacked by carbon monoxide.

The addition of water vapor increases the partial oxygen pressure in both CO and CO_2, whereas the oxidation rates are slightly lower compared to oxygen.

The oxidation rates reported in various literature references are compiled in Table 38.

Also [131] confirms that the oxidation in a CO_2 atmosphere follows a parabolic time law and that the oxidation rates are one order of magnitude lower than in pure oxygen.

Material (references)	Atmosphere	Temperature K (°C)	Oxidation rates cm^2/s	Oxidation rates $g^2/cm^4\,h$
RBSiC [132]	CO_2 1 bar	1473 (1200)	$9.1 \cdot 10^{-15}$	
	CO_2 0,5 bar	1273 (1000)	$4.6 \cdot 10^{-16}$	
		1473 (1200)	$6.6 \cdot 10^{-15}$	
		1573 (1300)	$1.6 \cdot 10^{-14}$	
SiC coated [133]	CO_2 1 bar	1473 (1200)	$3.0 \cdot 10^{-15}$	
	CO_2 0,5 bar	1273 (1000)	$6.5 \cdot 10^{-16}$	
		1473 (1200)	$2.7 \cdot 10^{-15}$	
		1573 (1300)	$4.1 \cdot 10^{-15}$	
SiC coated (Graphite) [134, 135]	CO_2	1623 (1350)	$3.7 \cdot 10^{-15}$	
	CO/50 % CO_2	1573 (1300)	$1.1 \cdot 10^{-15}$	
	CO/25 % CO_2	1523 (1250)	$6.6 \cdot 10^{-16}$	
		1623 (1350)	$3.2 \cdot 10^{-15}$	
	CO/15 % CO_2	1423 (1150)	$1.1 \cdot 10^{-16}$	
		1523 (1250)	$3.6 \cdot 10^{-15}$	
		1632 (1350)	$1.0 \cdot 10^{-14}$	
	CO	1673 (1400)	$1.4 \cdot 10^{-12}$	
Powder [136]	CO_2	1473 (1200)		$3.6 \cdot 10^{-12}$
		1573 (1300)		$1.4 \cdot 10^{-11}$
		1673 (1400)		$4.1 \cdot 10^{-11}$
		1873 (1600)		$3.2 \cdot 10^{-10}$
	N_2/17 % CO_2	1473 (1200)		$1.4 \cdot 10^{-12}$
		1573 (1300)		$6.7 \cdot 10^{-12}$
		1673 (1400)		$3.1 \cdot 10^{-11}$
		1773 (1500)		$5.6 \cdot 10^{-11}$
	N_2/2,5 % CO_2	1473 (1200)		$5.3 \cdot 10^{-13}$
		1573 (1300)		$4.4 \cdot 10^{-12}$
		1673 (1400)		$3.9 \cdot 10^{-11}$
		1773 (1500)		$9.6 \cdot 10^{-11}$
RBSiC [137]	CO_2 3 min	2370 (2097)		4.1 g/cm² h
	0,12 min	2720 (2447)		4.5 g/cm² h

Table 38: Oxidation rates of SiC in CO/CO_2 [130]

As the summarizing report in [138] shows, the oxidation resistance of SiC ceramics in salt melts, in water vapor or in reduced atmospheres can be improved by coating with mullite ($3Al_2O_3 \cdot 2SiO_2$) or with yttrium-stabilized zirconium oxide ($ZrO_2 \cdot Y_2O_3$) or with aluminium oxide (Al_2O_3). The layers can be applied in a plasma spraying process. Compared to other layers, mullite layers are advantageous in that their thermal expansion behavior is similar to that of the substrate.

The 9% Cr steels used in reactors cooled with CO_2 gas up to temperatures of 718 K (445°C) and the stainless steels of grade AISI 310, 316, 321 and 347 used up to temperatures of 973 K (700°C) are not only subjected to stresses regarding their oxidation behavior but also regarding their wear behavior by the gas flow. One option to reduce damage as a result of such stresses is to apply thin layers of ceramic materials combining high oxidation resistance with a high hardness.

The investigations described in [139, 140] dealt with the behavior of thin layers of titanium nitride or silicon nitride on pin- and disc-shaped specimens of the stainless steel AISI 321 (comparable with DIN Mat. No. 1.4541) in CO_2-containing atmospheres at 823 K (550°C) and at 973 K (700°C) with and without wear stresses. The type of the examined layers is summarized in Table 39.

No.	Type of layer	Layer thickness µm	Coating method
1	TiN	3	PVD vacuum arc
2	TiN	3	PVD sputtering
3	TiN	5	PVD diffusion
4	Si_3N_4	2	PVD vacuum

Table 39: Type, thickness and application method of the examined layers [139]

Examinations were performed in a simulated cooling gas consisting of CO_2 with 1% CO and 200 to 600 ppm H_2O at a pressure of 3.1 MN/m^2. The pin- or disc-shaped specimens were exposed for a period together of more than 10 000 h and the mass changes were determined after every 1000 to 2500 h. The results obtained for uncoated specimens and specimens with different coatings are plotted against both test temperatures in Figure 69 and in Figure 70.

The uncoated steel AISI 321 contains sufficient chromium to form a slowly growing chromium-rich cover layer during the oxidation in carbon dioxide at 823 or 973 K (550°C or 700°C). Defects in this cover layer facilitate the formation of rapidly growing iron oxide-rich needles, causing a noticeable increase in the oxidation rate.

The thin titanium nitride layers offer a hardly effective protection against oxidation under the conditions prevailing here. The growth rate at a test temperature of 823 K (550°C) shows that the major part of the layer was consumed after a test duration of 10 000 h. Hence, the titanium nitride layer is not able to prevent the oxidation of the substrate through the formation of iron oxide-rich needles. At the higher temperature of 973 K (700°C) the oxidation rate of the titanium nitride layer is further noticeably increased.

In contrast, the thin silicon nitride layer offers a clearly better protection of the substrate against oxidation in the CO_2-rich atmosphere. However, adhesion between the layer and the steel is not very good and hence spalling occurred during alternating thermal loading.

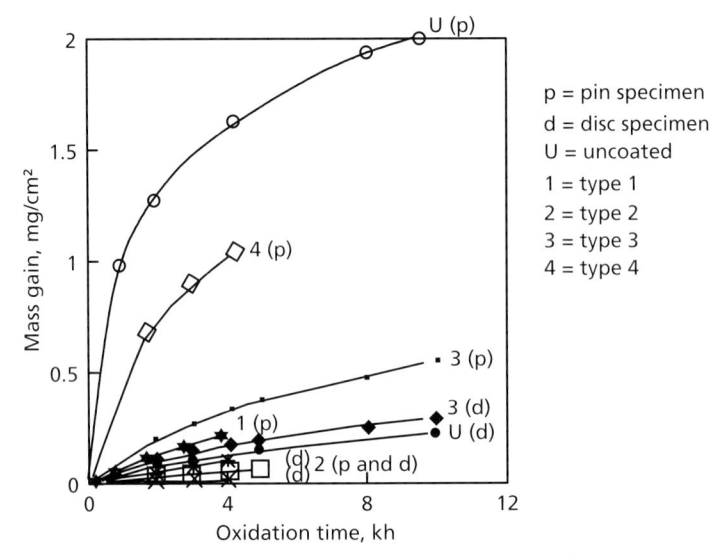

Figure 69: Mass gain of the specimens as a function of the test duration at 823 K (550°C) [139]

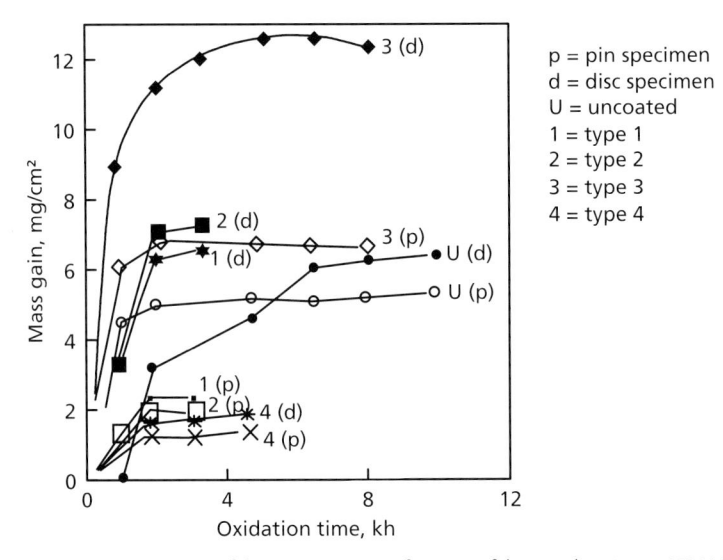

Figure 70: Mass gain of the specimens as a function of the test duration at 973 K (700°C) [139]

Of the ceramic materials, silicon carbide (SiC), utilized among other applications for highly stressed bearings and sealing rings, has an extremely broad scope of chemical resistance. For a SiC grade produced from sub-micron SiC particles by sintering without application of pressure (trade name: Hexoloy®) is reported to be resistant against aqueous carbon dioxide solutions or moist carbon dioxide gas up to about 473 K (200°C) [141].

The resistance of metal-ceramic materials to dry carbon dioxide gas at very high temperatures is indicated in Table 40.

Ceramic materials	Temperature K (°C)	Resistance	References
Silicon carbide	1273 (1000)	+	[117]
Boron carbide	1373 (1100)	+	[117]
Titanium carbide	1473 (1200)	−	[117]
Titanium nitride	1473 (1200)	⊕	[117]
Molybdenum disilicide	1873 (1600)	+	[117]
Cerium sulfide	2273 (2000)	+	[117]

+ = resistant   ⊕ = moderately resistant   − = not resistant

Table 40: Resistance of metal-ceramics to carbon dioxide gas

C
Organic materials / plastics

The abbreviated designations for plastics are used according to DIN EN ISO 1043-1 [41] and for rubbers according to DIN ISO 1629 [42].

Thermoplastics

Polyolefines and polyvinyl clorides

Thermoplastic (Abbreviation)	Medium	Concentration	Temperature K (°C)	Resistance	References
Polyethylene, low density (PE-LD)	aqueous CO_2 solution	saturated n. d.	293–333 (20–60) 293–333 (20–60)	+ +	[142] [143]
	CO_2 gas, moist	any n. d.	293–333 (20–60) 293–333 (20–60)	+ +	[13, 142] [143, 144]
	CO_2 gas, dry	any technically pure	293–333 (20–60) 293–333 (20–60)	+ +	[142] [13, 144]
Polyethylene, high density (PE-HD)	aqueous CO_2 solution	any saturated n. d. n. d.	293–333 (20–60) 293–333 (20–60) 294 (21) 293–333 (20–60)	+ + + +	[142, 145] [142, 146–148] [149] [143, 150–152]
	CO_2 gas, moist	any n. d.	293–333 (20–60) 293–333 (20–60)	+ +	[13, 142] [143, 144, 146, 147, 151, 153]
	CO_2 gas, dry	any technically pure	293–333 (20–60) 293 333 (20–60)	+ +	[14, 22, 142, 154] [13, 143–148, 150–153, 155, 156]

\+ = resistant   ⊕ = moderately resistant   – = not resistant
RT = room temperature
n. d. = not defined
[1] Test duration: 60 d
[2] Test duration: 90 d

Table 41: Resistance of polyolefines and polyvinyl chlorides to aqueous carbon dioxide solutions and carbon dioxide gas

Table 41: Continued

Thermoplastic (Abbreviation)	Medium	Concentration	Temperature K (°C)	Resistance	References
Polyethylene, high molar mass (PE-HMW)	aqueous CO_2 solution	any	293–333 (20–60)	+[1]	[157]
	CO_2 gas	technically pure	293–333 (20–60)	+[1]	[157]
Polyethylene, very high molar mass (PE-UHMW)	aqueous CO_2 solution	any	293–333 (20–60)	+[1]	[157]
		n. d.	RT	+	[149, 158]
	CO_2 gas	technically pure	RT	+	[158–160]
		technically pure	293–333 (20–60)	+[1]	[157]
Polyethylene, cross-linked (PE-X)	aqueous CO_2 solution	saturated	≤ 333 (≤ 60)	+	[147]
		n. d.	RT	+	[158]
	CO_2 gas, moist	n. d.	293–333 (20–60)	+	[147, 161]
	CO_2 gas, dry	technically pure	RT	+	[158, 160]
		technically pure	293–333 (20–60)	+	[147, 161]
Polypropylene (PP)	aqueous CO_2 solution	any	293–333 (20–60)	+	[145, 162]
		saturated	293–333 (20–60)	+	[147, 148, 163]
		saturated	353 (80)	+	[146]
		saturated	353 (80)	+ to ⊕	[163]
		n. d.	294 (21)	+	[149]
		n. d.	293–333 (20–60)	+	[142, 152, 164]
		n. d.	293–353 (20–80)	+	[150, 165]
	CO_2 gas, moist	any	293–333 (20–60)	+	[13]
		n. d.	293–333 (20–60)	+	[142–144, 147, 153, 164]
		n. d.	≤ 339 (≤ 66)	+	[166, 167]
		n. d.	293–353 (20–80)	+	[146, 163]
	CO_2 gas, dry	any	293–333 (20–60)	+	[14, 22, 154]
		technically pure	293–333 (20–60)	+	[13, 142–145, 147, 148, 152, 162]
		technically pure	≤ 339 (≤ 66)	+	[165–167]
		technically pure	293–353 (20–80)	+	[146, 150, 153, 156, 163]

+ = resistant   ⊕ = moderately resistant   – = not resistant
RT = room temperature
n. d. = not defined
[1] Test duration: 60 d
[2] Test duration: 90 d

Table 41: Resistance of polyolefines and polyvinyl chlorides to aqueous carbon dioxide solutions and carbon dioxide gas

Table 41: Continued

Thermoplastic (Abbreviation)	Medium	Concentration	Temperature K (°C)	Resistance	References
Polybutene (PB)	aqueous CO_2 solution	saturated	293–333 (20–60)	+	[144]
	CO_2 gas, moist	n. d.	293–333 (20–60)	+	[144, 147]
	CO_2 gas, dry	technically pure	293–333 (20–60)	+	[144]
Polyisobutylene (PIB)	aqueous CO_2 solution	saturated	293 (20)	–	[13]
	CO_2 gas, moist	any	313–373 (40–100)	+	[13]
	CO_2 gas, dry	technically pure	333 (60)	+	[13]
Polymethylpentene	aqueous CO_2 solution	n. d.	293–333 (20–60)	+[2]	[142]
	CO_2 gas	technically pure	293–333 (20–60)	+[2]	[142]
Polyvinyl chloride, without plasticizers (PVC-U)	aqueous CO_2 solution	≤ saturated	293–333 (20–60)	+	[168]
		saturated	293 (20)	+	[13]
		saturated	293–313 (20–40)	+	[169]
		saturated	293–333 (20–60)	+	[144, 146–148, 163, 166, 167]
		saturated	333 (60)	⊕	[169]
		n. d.	298 (25)	+	[152]
		n. d.	293–333 (20–60)	+	[142, 150, 165]
		n. d.	333 (60)	⊕	[152]
	CO_2 gas, moist	any	313 (40)	+	[13]
		any	333 (60)	⊕	[13]
		n. d.	293–313 (20–40)	+	[153]
		n. d.	293–333 (20–60)	+	[144, 146, 147, 163, 166, 167]
	CO_2 gas, dry	n. d.	333 (60)	⊕	[153]
		any	293–333 (20–60)	+	[14, 22, 154]
		technically pure	293–333 (20–60)	+	[13, 142, 144–148, 150, 152, 153, 156, 163, 165–167, 169]

+ = resistant ⊕ = moderately resistant – = not resistant
RT = room temperature
n. d. = not defined
[1] Test duration: 60 d
[2] Test duration: 90 d

Table 41: Resistance of polyolefines and polyvinyl chlorides to aqueous carbon dioxide solutions and carbon dioxide gas

Table 41: Continued

Thermoplastic (Abbreviation)	Medium	Concentration	Temperature K (°C)	Resistance	References
Polyvinyl chloride, chlorinated (PVC-C)	aqueous CO_2 solution	≤ saturated	293–353 (20–80)	+	[170]
		saturated	293–333 (20–60)	+	[144, 163]
		saturated	≤ 353–358 (≤ 80–85)	+	[146, 147, 166, 167]
		saturated	293–353 (20–80)	+ to ⊕	[163]
		n. d.	296 (23)	+	[142]
		n. d.	293–353 (20–80)	+	[150, 171]
		n. d.	296–366 (23–93)	+	[172]
	CO_2 gas, moist	n. d.	293–333 (20–60)	+	[144]
		n. d.	293–353 (20–80)	+	[146, 147, 153, 163]
		n. d.	≤358 (≤85)	+	[166, 167]
	CO_2 gas, dry	technically pure	293–333 (20–60)	+	[144]
		technically pure	293–353 (20–80)	+	[146, 147, 150, 153, 156, 163, 170, 171]
		technically pure	≤ 358 (≤ 85)	+	[166, 167]
		technically pure	296–366 (23–93)	+	[172]
Polyvinyl chloride, with plasticizer (PVC-P)	aqueous CO_2 solution	saturated	293–313 (20–40)	+	[148]
	CO_2 gas, dry	technically pure	293–333 (20–60)	+	[148]
	CO_2 gas	any	293–313 (20–40)	+	[14]
		technically pure	296 (23)	+	[173]
Polyvinylidene chloride (PVDC)	CO_2 gas	technically pure	296–325 (23–52)	+	[142]

+ = resistant   ⊕ = moderately resistant   – = not resistant
RT = room temperature
n. d. = not defined
[1] Test duration: 60 d
[2] Test duration: 90 d

Table 41: Resistance of polyolefines and polyvinyl chlorides to aqueous carbon dioxide solutions and carbon dioxide gas

The thermoplastics PE-HD, PP and PVC-U which are frequently used in the manufacture of chemical equipment, are suited for applications with aqueous carbon dioxide solutions and moist or dry carbon dioxide gas up to 333 K (60°C) (PE-HD), 333 to 353 K (60 to 80°C) (PP) and 313 to 333 K (40 to 60°C) (PVC-U). The upper application temperature limit for PVC-C is 333 to 353 K (60 to 80°C).

In [21] the dependence of the permeability coefficient is investigated for carbon dioxide and other gases of a co-extruded low-density foil of linear polyethylen

(PE-LLD) on the temperature and a stretching treatment. PE-LLD containing about 8 mol% 1-octene as comonomer exhibited a crystallinity of about 30 %; the three-layer foil (total thickness 23 μm) was examined untreated and after stretching in longitudinal and transverse direction in relation to extrusion direction. Figure 71 shows that the unstretched foil exhibited a slightly higher permeability for carbon dioxide in the temperature range investigated compared to the two stretched foils; in general, however, the dependence of the permeability coefficient on the type of after-treatment was rather low. The dependence on the temperature was much more marked; in Figure 71 two different temperature ranges can be seen, i.e. a lower range of about 298 to 328 K (25 to 55°C) and an upper range of > 328 to 358 K (> 55 to 85°C), with a linearity each according to the Arrhenius relation. Using the Arrhenius equation an activation energy of 30.5 kJ/mol was determined for the lower range and of 17.4 kJ/mol for the upper range of the unstretched foil. The different activation energies were attributed to a temperature-induced change in the microstructure of PE-LLD.

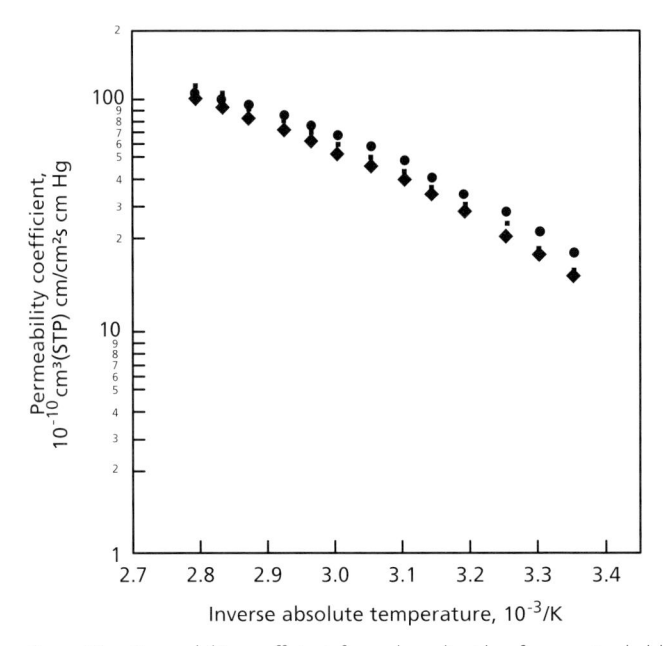

Figure 71: Permeability coefficient for carbon dioxide of a co-extruded low-density foil of linear polyethylene (PE-LLD) [21]
● unstretched foil; ◆ foil stretched in longitudinal direction; ■ foil stretched in transverse direction

The dependence of the permeability coefficient for carbon dioxide, oxygen and nitrogen in a poly-4-methylpentene-1 membrane on the applied gas pressure was examined in the temperature range from 293 to 318 K (20 to 45°C) in [174]. Whereas no dependence on the gas pressure was observed for oxygen and nitrogen, the permeability coefficient for carbon dioxide noticeably increased with a rising gas pressure. It can be seen from Figure 3 that in the temperature range from 303 to 318 K

(30 to 45°C) there is a linear relationship between the logarithm of the mean permeability coefficient and the gas pressure, attributed to the growing plasticization of the polymer by sorbed carbon dioxide with increasing gas pressure. It was learned from the sorption measurements that the glass transition temperature of the polymer of 307 K (34°C) was lowered by sorbed carbon dioxide such that the polymer assumed a rubber-like condition from about 303 K (30°C) and the there was a linear relation between the logarithm of the mean permeability coefficient and the gas pressure from this temperature.

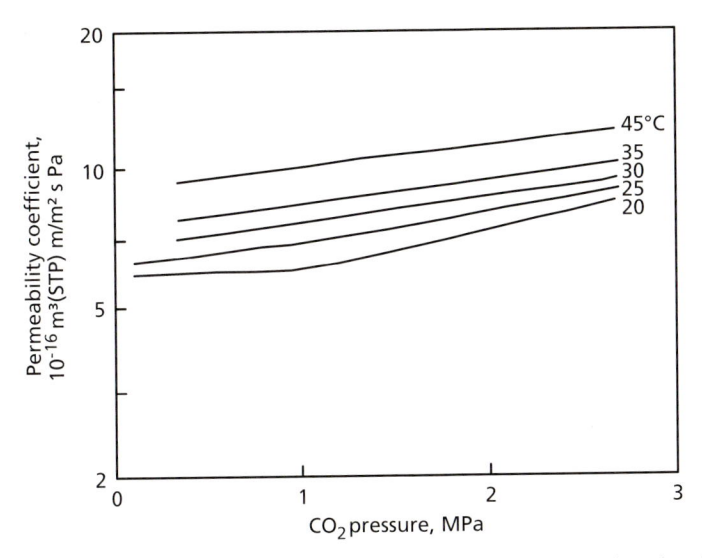

Figure 72: Dependence of the mean permeability coefficient for carbon dioxide on the CO_2 pressure of a poly-4-methylpentene-1 membrane [174]

The dependence of the permeability coefficient for carbon dioxide and other gases of a plasticizing polyvinyl chloride (PVC-P) on the plasticizer content and the addition of a second polymer was examined in [175]. Di(2-ethylhexyl)-phthalate was used as the plasticizer and a graft copolymer of 50% vinyl chloride and 50% ethylene vinyl acetate copolymer or acrylonitrile-butadiene rubber (NBR) was used as the second polymer. The composition of the polymer mixtures and the permeability coefficients determined at 27% and 65% relative air humidity are indicated in Table 42. A reduction of the plasticizer content of about 33% (specimen 1) to about 21% (specimen 2) lead to a strong reduction of the permeability coefficient. A comparably strong reduction of the permeability coefficients could also be reached by adding a fairly small amount of a graft copolymer (specimen 3) or the rubber (specimen 5); here, the plasticizer content decreased only moderately. Further reduction of the permeability coefficient was possible by using a clearly increased content of graft copolymer (specimen 4) or a mixture of all three polymers (specimen 6); however, it must be assumed that such a marked modification of the composition probably also causes a change in the mechanical and other physical properties not examined here.

No.	Composition in % by weight					Permeability coefficient
	PVC	Graft copolymer[1]	NBR	Plasticizer[2]	Additives	$10^{-13} \times \frac{cm^3 \, cm}{cm^2 \, s \, Pa}$
1	59.5	–	–	32.7	7.8	120.8
2	69.8	–	–	20.9	9.3	63.2
3	53.6	8.5	–	29.8	8.1	73.1
4	17.9	59.2	–	12.7	10.2	52.1
5	53.6	–	10.0	29.4	7.0	66.0
6	19.8	28.2	33.3	12.3	6.4	60.1

Measurement temperature: 300 K (27°C); Relative humidity: 65 %
[1] Graft copolymer of vinyl chloride and ethylene vinyl acetate copolymer
[2] Di(2-ethylhexyl)-phthalate

Table 42: Permeability coefficient for carbon dioxide of plasticizer-containing polyvinyl chloride (PVC-P) and mixtures with other polymers [175]

In a polyvinyl chloride (suspension PVC, K value 66 to 69) plasticized with 40 phr di(2-ethylhexyl)-phthalate, the plasticizer was partially replaced by 7.5 to 25 phr of an acrylonitrile-butadiene rubber (NBR) with an acrylonitrile content of 34 % to reduce the permeability for carbon dioxide and other gases and the extractability of the plasticizer. The composition of the test parameters is indicated in Table 43. The permeability was determined for specimens with a thickness of 0.3 mm at 283, 298 and 313 K (10, 25 and 40°C). As can be seen from Figure 73, the permeability coefficient for carbon dioxide strongly increased in all specimens as the temperature increased and that at the same temperature the permeability coefficient continuously decreased up to an NBR content of 15 phr and further increasing the NBR content did not cause any further noticeable reduction. Up to an NBR addition of 15 phr the density and the hardness (Shore A) were only slightly influenced [176].

Components	Composition in g			
Designation	PVC-P	NBR-7.5	NBR-15	NBR-25
Polyvinyl chloride	100	100	100	100
Acrylonitrile butadiene rubber (NBR)	–	7.5	15	25
Di(2-ethylhexyl)-phthalate	40	32.5	25	15
Dioctyl adipate	10	10	10	10
Epoxidized oil	7	7	7	7
Additives	3	3	3	3

Table 43: Composition of the plasticizer-containing polyvinyl chloride specimens [176]

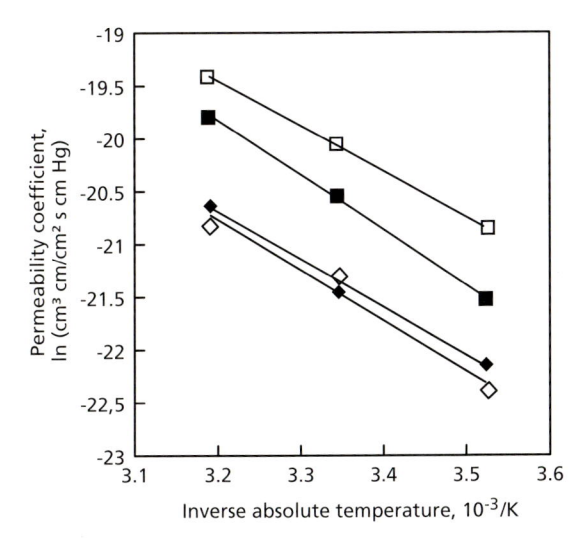

Figure 73: Dependence of the permeability coefficient for carbon dioxide on the temperature and the NBR addition of plasticizer-containing polyvinyl chloride (PVC-P) [176]
□ without NBR; ■ with 7.5 phr NBR; ◇ with 15 phr NBR; ◆ with 25 phr NBR

Fluoropolymers and high-temperature thermoplastics

Polymer	Medium	Concentration	Temperature K (°C)	Resistance	References
a) Fluoropolymers					
Polytetrafluorethylene (PTFE)	aqueous CO_2 solution	saturated	293–353 (20–80)	+	[177]
		saturated	293–393 (20–120)	+	[146, 163]
		saturated	≤ 450 (≤ 177)	+	[166, 167]
		n. d.	≤ 450 (≤ 177)	+	[178]
	CO_2 gas, moist	n. d.	293–353 (20–80)	+	[177]
		n. d.	293–393 (20–120)	+	[146, 163]
		n. d.	≤ 477 (≤ 204)	+	[166, 167, 178]
	CO_2 gas, dry	technically pure	293–353 (20–80)	+	[177]
		technically pure	293–373 (20–100)	+	[156]
		technically pure	293–393 (20–120)	+	[146, 163]
		technically pure	≤ 477 (≤ 204)	+	[166, 167]

+ = resistant   ⊕ = moderately resistant   – = not resistant
RT = room temperature
n. d. = not defined

Table 44: Resistance of fluoropolymers and high-temperature thermoplastics to aqueous carbon dioxide solutions and carbon dioxide gas

Table 44: Continued

Polymer	Medium	Concentration	Temperature K (°C)	Resistance	References
Tetrafluoroethylene-perfluoropropylvinyl-ether copolymer (PFA)	CO_2 gas, moist	any n. d.	≤ 533 (≤ 260) 298–423 (25–150)	+ +	[179] [180]
	CO_2 gas, dry	any technically pure	≤ 533 (≤ 260) 298–423 (25–150)	+ +	[179] [180]
Tetrafluoroethylene-hexafluoropropylene copolymer (FEP)	CO_2 gas, moist	any n. d.	≤ 478 (≤ 205) 298–423 (25–150)	+ +	[181] [180]
	CO_2 gas, dry	any technically pure	≤ 478 (≤ 205) 298–423 (25–150)	+ +	[181] [180]
	CO_2 gas	technically pure	RT	+	[159, 160]
Tetrafluoroethylene/perfluoromethyl vinyl ether copolymer (MFA)	CO_2 gas, moist	n. d.	298–423 (25–150)	+	[180]
	CO_2 gas, dry	technically pure	298–423 (25–150)	+	[180]
Ethylene chlorotri-fluoroethylene copolymer (ECTFE)	aqueous CO_2 solution	saturated n. d.	≤ 393 (≤ 120) 296–422 (23–149)	+ +	[146] [182]
	CO_2 gas, moist	n. d.	≤ 393–398 (≤ 120–125)	+	[146, 180]
		n. d.	296–422 (23–149)	+	[142, 182]
	CO_2 gas, dry	100%	293–393 (20–120)	+	[22, 146]
		technically pure	296–422 (23–149)	+	[142, 180, 182]
Ethylene tetrafluoro-ethylene copolymer (ETFE)	aqueous CO_2 solution	n. d.	≤ 423 (≤ 150)	+	[142, 183]
	CO_2 gas, moist	n. d.	≤ 423 (≤ 150)	+	[142, 183]
	CO_2 gas, dry	technically pure	≤ 423 (≤ 150)	+	[142, 183]

+ = resistant   ⊕ = moderately resistant   – = not resistant
RT = room temperature
n. d. = not defined

Table 44: Resistance of fluoropolymers and high-temperature thermoplastics to aqueous carbon dioxide solutions and carbon dioxide gas

Table 44: Continued

Polymer	Medium	Concentration	Temperature K (°C)	Resistance	References
Polyvinylidene fluoride (PVDF)	aqueous CO_2 solution	≤ saturated	≤ 373 (≤ 100)	+	[168, 184]
		saturated	293–353 (20–80)	+	[22]
		saturated	293–373 (20–100)	+	[148]
		saturated	293–393 (20–120)	+	[146, 163]
		saturated	≤ 411 (≤ 138)	+	[166, 167]
		n. d.	293–353 (20–80)	+	[150]
		n. d.	≤ 408 (≤ 135)	+	[142, 185]
	CO_2 gas, moist	n. d.	293–353 (20–80)	+	[153]
		n. d.	293–393 (20–120)	+	[146, 163]
		n. d.	298–398 (25–125)	+	[180]
		n. d.	≤ 411 (≤ 138)	+	[166, 167]
	CO_2 gas, dry	any	293–393 (20–120)	+	[22]
		any	≤ 383–413 (≤ 110–140)	+	[184]
		technically pure	293–373 (20–100)	+	[147, 150, 153]
		technically pure	293–393 (20–120)	+	[144, 146, 148, 156, 163]
		technically pure	298–398 (25–125)	+	[180, 186]
		technically pure	≤ 413 (≤ 140)	+	[142, 185]
b) High-temperature thermoplastics					
Polyphenylene sulfide (PPS)	aqueous CO_2 solution	n. d.	≤ 366 (≤ 93)	+	[142, 187]
	CO_2 gas	technically pure	RT	+	[188]
		technically pure	≤ 366 (≤ 93)	+	[142, 187]
Polysulfone (PSU)	CO_2 gas	technically pure	RT	+	[188]
		technically pure	333 (60)	+	[156]
Polyethersulfone (PESU)	CO_2 gas	n. d.	293 (20)	–	[142]

+ = resistant ⊕ = moderately resistant – = not resistant
RT = room temperature
n. d. = not defined

Table 44: Resistance of fluoropolymers and high-temperature thermoplastics to aqueous carbon dioxide solutions and carbon dioxide gas

Table 44: Continued

Polymer	Medium	Concentration	Temperature K (°C)	Resistance	References
Polyetheretherketone (PEEK)	aqueous CO_2 solution	saturated saturated	RT ≤ 408 (≤ 135)	+ +	[188] [166, 167, 189]
	CO_2 gas, moist	n. d. n. d.	< 408 (< 135) ≤ 561 (≤ 288)	+ to ⊕ +	[189] [166, 167]
	CO_2 gas, dry	technically pure technically pure technically pure	RT < 408 (< 135) ≤ 561 (≤ 288)	+ + +	[188, 190] [189] [166, 167]
Polyimide (PI)	CO_2 gas	technically pure	296–338 (23–65)	+ to ⊕	[142]
Polyetherimide (PEI)	CO_2 gas	technically pure	RT	+	[145]
Polyamideimide (PAI)	CO_2 gas	technically pure	RT	+	[188]

+ = resistant   ⊕ = moderately resistant   − = not resistant
RT = room temperature
n. d. = not defined

Table 44: Resistance of fluoropolymers and high-temperature thermoplastics to aqueous carbon dioxide solutions and carbon dioxide gas

The fully fluorinated polymers PTFE, PFA, FEP and MFA, if exposed to aqueous carbon dioxide solutions or moist carbon dioxide gas, can be used up to about 423 K (150°C) and, if exposed to dry carbon dioxide gas, up to about 473 K (200°C). The upper temperature limit for the use of the partially fluorinated polymers ECTFE, ETFE and PVDF is 393 to 423 K (120 to 150°C) (ECTFE and ETFE) or 373 to 393 K (100 to 120°C) (PVDF). The upper temperature limit for the use of the PPS belonging to the group of the high-temperature thermoplastics is 363 to 373 K (90 to 100°C).

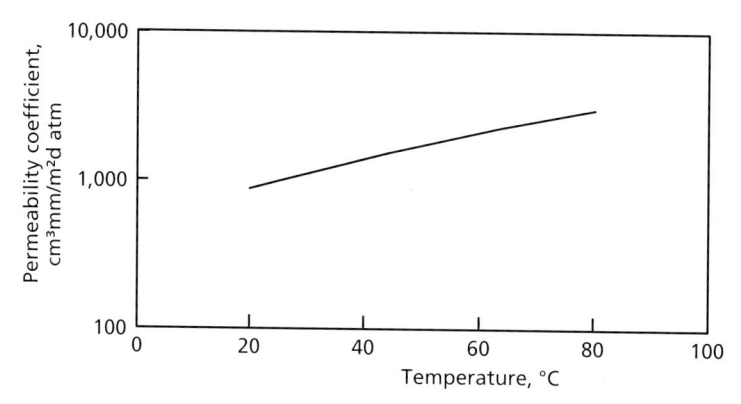

Figure 74: Dependence of the permeability coefficient for carbon dioxide of tetrafluoroethylene perfluoropropylvinylether copolymer (PFA) on temperature [191]

The permeability of the fluorothermoplastics for carbon dioxide gas strongly increases in the temperature range from 293 to 373 K (20 to 100°C). The dependence of the permeability coefficient for carbon dioxide gas is depicted in Figure 74 for PFA und in Figure 75 for PVDF (homopolymer).

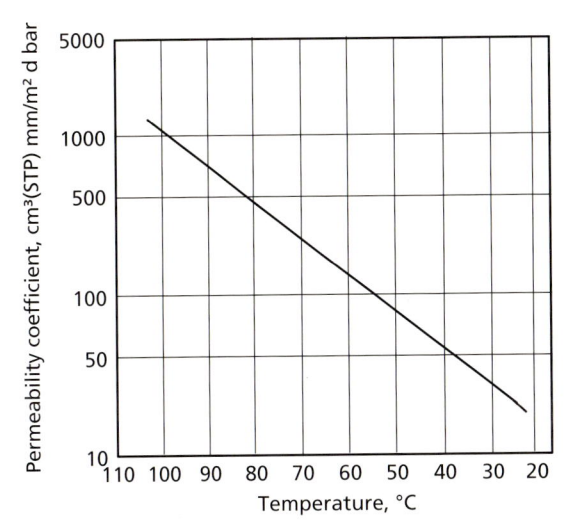

Figure 75: Dependence of the permeability coefficient for carbon dioxide on polyvinylidene fluoride (PVDF) on temperature [24]

Further thermoplastics

Further thermoplastics as listed in Table 45 are resistant to aqueous carbon dioxide solutions or carbon dioxide gas (moist or dry) up to about 323 to 333 K (50 to 60°C). However, if exposed to carbon dioxide gas the plasticizing effect needs to be considered, occurring first of all in polymers with carbonyl or ether groups. Several examples are given below.

Polymer	Medium	Concentration	Temperature K (°C)	Resistance	References
Polystyrene (PS)	CO_2 gas	any	293–323 (20–50)	+	[14]
		technically pure	293–323 (20–50)	+	[13, 142, 145, 192]
		technically pure	293–333 (20–60)	+	[145]
Polystyrene, shock-resistant (PS-HI)	CO_2 gas	technically pure	293–323 (20–50)	+	[142, 192]

+ = resistant − = not resistant
RT = room temperature
n. d. = not defined

Table 45: Resistance of further thermoplastics to aqueous carbon dioxide solutions and carbon dioxide gas

Table 45: Continued

Polymer	Medium	Concentration	Temperature K (°C)	Resistance	References
Styrene-butadiene copolymer (SB)	CO_2 gas	technically pure	293–323 (20–50)	+	[13]
		technically pure	RT	+	[145]
Acrylonitrile-butadiene-styrene copolymer (ABS)	aqueous CO_2 solution	saturated	293–323 (20–50)	+	[144]
		saturated	≤ 358 (≤ 85)	+	[147]
	CO_2 gas, moist	n. d.	≤ 311 (≤ 38)	+	[166, 167]
		n. d.	293–323 (20–50)	+	[144]
		n. d.	293–333 (20–60)	+	[153]
		n. d.	≤ 344 (≤ 71)	+	[147]
	CO_2 gas, dry	technically pure	≤ 311 (≤ 38)	+	[166, 167]
		technically pure	293–323 (20–50)	+	[13, 144, 193]
		technically pure	293–333 (20–60)	+	[145, 153]
		technically pure	≤ 344 (≤ 71)	+	[147]
Styrene-acrylonitrile copolymer (SAN)	CO_2 gas	technically pure	293–323 (20–50)	+	[13, 142, 193]
		technically pure	RT–333 (RT–60)	+	[145]
Methyl-methacrylate-acrylonitrile-butadiene-styrene copolymer (MABS)	CO_2 gas	technically pure	293–323 (20–50)	+	[193]
Acrylate styrene acrylonitrile copolymer (ASA)	CO_2 gas	technically pure	293–323 (20–50)	+	[13, 142, 193]
		technically pure	RT-333 (RT-60)	+	[145]
Polyethylene terephthalate (PET)	aqueous CO_2 solution	any	343 (70)	+	[13]
		saturated	RT	+	[188]
	CO_2 gas	any	343 (70)	+	[13]
		technically pure	RT	+	[188]
		technically pure	293 (20)	–	[142]
	CO_2 (200 bar)	technically pure	RT	+	[13]
Polybutylene terephthalate (PBT)	aqueous CO_2 solution	any	343 (70)	+	[13]
	CO_2 gas	any	293–333 (20–60)	+	[194]
		any	343 (70)	+	[13]
		technically pure	343 (70)	+	[31]
	CO_2 (200 bar)	technically pure	RT	+	[13]
		technically pure	296–343 (23–70)	+	[142]

+ = resistant   – = not resistant
RT = room temperature
n. d. = not defined

Table 45: Resistance of further thermoplastics to aqueous carbon dioxide solutions and carbon dioxide gas

Table 45: Continued

Polymer	Medium	Concentration	Temperature K (°C)	Resistance	References
Polycarbonate (PC)	aqueous CO_2 solution	any	343 (70)	+	[13]
	CO_2 gas, moist	n. d.	RT	+	[195, 196]
	CO_2 gas	any technically pure	343 (70) RT	+ +	[13] [145]
	CO_2 (200 bar)	technically pure	RT	+	[13]
Polyamide 6 (PA6)	aqueous CO_2 solution	any	343 (70)	+	[13]
	CO_2 gas	any technically pure	343 (70) 343 (70)	+ +	[13] [31, 197]
	CO_2 (200 bar)	technically pure technically pure	RT 296–343 (23–70)	+ +	[13] [142]
Polyamide 66 (PA66)	aqueous CO_2 solution	any saturated	343 (70) RT	+ +	[13] [142, 198]
	CO_2 gas	any technically pure	343 (70) 343 (70)	+ +	[13] [31, 197]
	CO_2 (200 bar)	technically pure	RT	+	[13]
Polyamide 610 (PA610)	aqueous CO_2 solution	any	343 (70)	+	[13]
	CO_2 gas	any technically pure	343 (70) 296–343 (23–70)	+ +	[13] [142]
	CO_2 (200 bar)	technically pure	RT	+	[13]
Polyamide 11 (PA11)	aqueous CO_2 solution	any n. d.	343 (70) 293–363 (20–90)	+ +	[13] [32]
	CO_2 gas	any	343 (70)	+	[13]
	CO_2 (200 bar)	technically pure	RT	+	[13]

+ = resistant   − = not resistant
RT = room temperature
n. d. = not defined

Table 45: Resistance of further thermoplastics to aqueous carbon dioxide solutions and carbon dioxide gas

Table 45: Continued

Polymer	Medium	Concentration	Temperature K (°C)	Resistance	References
Polyamide 12 (PA12)	aqueous CO_2 solution	any	343 (70)	+	[13]
	CO_2 gas	any	343 (70)	+	[13]
	CO_2 (200 bar)	technically pure	RT	+	[13]
Polyoxymethylene (POM)	aqueous CO_2 solution	any saturated	343 (70) RT	+ +	[13] [188]
	CO_2 gas	any technically pure technically pure	343 (70) RT 343 (70)	+ + +	[13] [188, 199] [31]
	CO_2 (200 bar)	technically pure technically pure	RT 296–343 (23–70)	+ +	[13] [142]
Polymethylmethacrylate (PMMA)	aqueous CO_2 solution	any	343 (70)	+	[13]
	CO_2 gas	any technically pure	343 (70) RT	+ +	[13] [145]
	CO_2 (200 bar)	technically pure	RT	+	[13]
Polyphenylene oxide (PPE)	aqueous CO_2 solution	saturated	293 (20)	+	[13]
	CO_2 gas, moist	any	313 (40)	+	[13]
Ethylene-vinyl acetate copolymer (EVA)	CO_2 gas, moist	any	293–333 (20–60)	+	[13]
	CO_2 gas, dry	technically pure	333 (60)	+	[13]

+ = resistant – = not resistant
RT = room temperature
n. d. = not defined

Table 45: Resistance of further thermoplastics to aqueous carbon dioxide solutions and carbon dioxide gas

In [200] the plasticizing effect of carbon dioxide gas was monitored using three thermoplastics, i.e. polystyrene (PS), polycarbonate (PC) and polymethylmethacrylate (PMMA) by means of mechanical measurements up to high CO_2 pressures (120 bar). The polymer specimens placed on a three-point bending test bench in a heatable pressure chamber (dimensions: 80 mm × 20 mm × 2–3 mm; distance between support points: 46 mm) were exposed to CO_2 pressures from 0 to 120 bar and heated with a heating rate of 1°C/min to 413–433 K (140–160°C). With an LVDT (linear variable displacement transducer) centrally arranged on the specimens the mechanical changes of the specimens were registered. The softening temperature of the outer layer of the specimens was defined as that temperature at which the displacement transducer indicated a displacement of 0.01 mm within a temperature interval of 10 K for the first time. The data in Table 46 shows that a drop of the softening temperature occurred as the CO_2 pressure increased, caused by the sorption of carbon dioxide in the outer specimen layer. Comparing the softening temperatures with the data reported in the literature as to the glass transition temperature of the three polymers when exposed to carbon dioxide within the reported pressure range, there is an acceptable qualitative correspondence.

CO_2 pressure bar	Softening temperature of the outer layer K (°C)		
	Polystyrene	Polycarbonate	Polymethylmethacrylate
0	350 (77)	422 (149)	363 (90)
20	340 (67)	403 (130)	362 (89)
54	313 (40)	390 (117)	343 (70)
85	313 (40)	370 (97)	347 (74)
120	311 (38)	313 (40)	353 (80)

Table 46: Change in the softening temperature of thermoplastics when exposed to carbon dioxide [200]

The permeability coefficients for carbon dioxide and other gases at 308 K (35°C) were determined for polymers of differently substituted styrene monomers and a correlation with the specific free volume of the polymers was shown. The individual polymers, their mean molar mass, their specific free volume and the permeability coefficient are indicated in Table 47. As can be seen from Figure 76, there is an almost linear dependence of the logarithm of the permeability coefficient on the reciprocal specific free volume of the polymers; the permeability coefficient increased if the specific free volume increased. Moreover, it was found that the permeability coefficient of poly-p-acetoxystyrene and poly-p-methoxystyrene depended on the applied gas pressure; with an increasing gas pressure a flat minimum was passed through at about 15 atm; when the applied maximum pressure of 30 bar was reduced, the permeability coefficient significantly increased. This behavior was attributed to the fairly good solubility of carbon dioxide in both polymers due to the carbonyl and ether groups and the related plasticizing effect [201].

Polymer	Abbreviation	Glass transition temperature K (°C)	Mean molar mass M_w kg/mol	Specific free volume cm^3/g	Permeability coefficient $10^{-13} \times cm^3(STP)$ cm/cm^2 s Pa
Poly-α-methyl-styrene	PαMS	438 (165)	135	0.135	2.25
Poly-p-chlorstyrene	PCS	399 (126)	650	0.124	3.23
Polystyrene	PS	375 (102)	350	0.169	9.30
Poly-p-acetoxy-styrene	PAS	399 (126)	250	0.141	12.2
Poly-p-fluorstyrene	PFS	377 (104)	150	0.146	12.9
Poly-p-methoxy-styrene	PMxS	378 (105)	980	0.146	14.2
Poly-p-methyl-styrene	PMS	381 (108)	357	0.176	22.4
Poly-p-tert-butyl-styrene	PtBS	411 (138)	310	0.207	105

STP = Standard temperature and pressure (273,15 K (0°C); 1,13 × 10^5 Pa)

Table 47: Permeability coefficient for carbon dioxide of substituted polystyrenes at 308 K (35°C) [201]

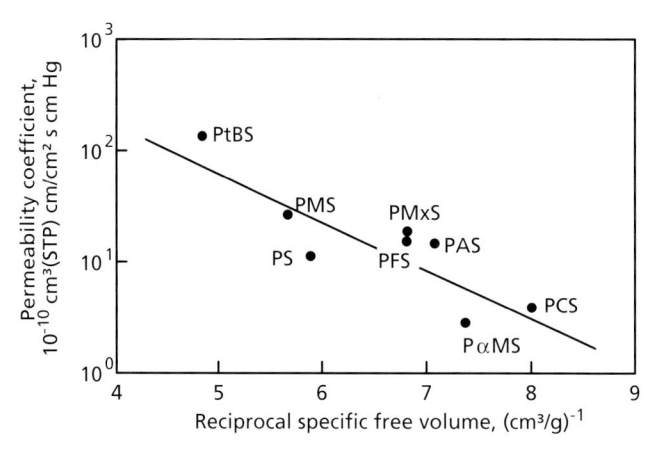

Figure 76: Correlation of the permeability coefficient for carbon dioxide at 308 K (35°C) with the specific free volumes of substituted polystyrenes [201]
(For abbreviations refer to Table 47)

The dependence of the permeability coefficients for carbon dioxide and other gases on the free volume fraction was examined at 308 K (35°C) using membranes of polymethylmethacrylate (PMMA) with a high molar mass produced in a casting process from PMMA solutions with solvents of a different polarity and different boiling points. The free volume fraction of the membranes calculated from the density measurements and the mean radius of the voids of the free volume determined by means of the positron annihilation lifetime spectroscopy are listed in Table 48; the smallest free volume fraction occurred with dichloromethane as a solvent, whereas the highest free volume fraction occurred with methyl isobutyl ketone. Figure 77 shows an almost linear increase in the logarithm of the permeability coefficient as the free volume fraction increased. Further investigations revealed that the pressure dependence of the PMMA membrane was especially marked with the highest free volume fraction [202].

Solvent	Density PMMA g/cm^3	Free volume fraction PMMA	Mean radius Free volume voids nm
Dichloromethane	1.135	0.173	0.283
Tetrahydrofuran	1.126	0.179	0.288
Ethyl acetate	1.124	0.181	0.292
Butyl acetate	1.119	0.184	0.297
Methyl isobutyl ketone	1.114	0.188	0.300

Mean molar mass M_w of the PMMA: 996 kg/mol
Solids content of the PMMA solutions: 14.3 Ma%

Table 48: Characterization of polymethylmethacrylate (PMMA) membranes produced from solutions in the casting process [202]

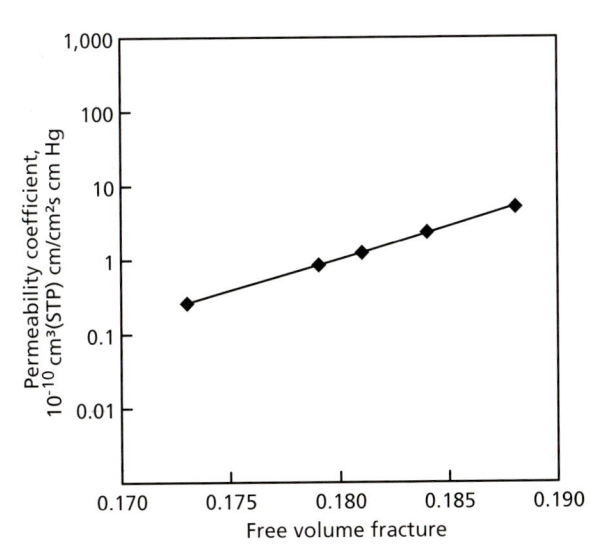

Figure 77: Dependence of the permeability coefficient for carbon dioxide on the free volume fraction of the polymethylmethacrylate (PMMA) membranes at 308 K (35°C) [202]

A particularly high dependence of the permeability coefficient on carbon dioxide pressure was also observed for polyethylmethacrylate. During measurements performed at 308 K (35°C) the permeability coefficient strongly increased when the carbon dioxide pressure increased (refer to Figure 78) since the polyethylmethacrylate was increasingly plasticized through the sorption of carbon dioxide, and hence diffusion was enhanced [34].

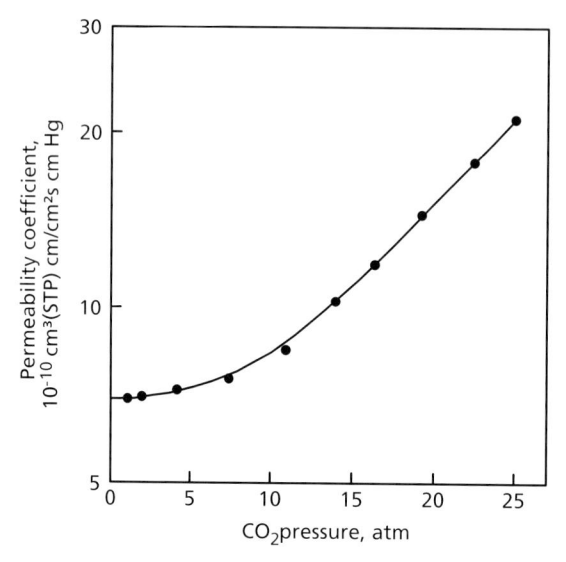

Figure 78: Permeability coefficient for carbon dioxide of polyethylmethacrylate at 308 K (35°C) as a function of the CO_2 pressure [34]

In a wide investigation of the plasticizing effect of carbon dioxide using 11 amorphous polymers with high glass transition temperatures (448 to 586 K (175 to 313°C)) it was established in [25] that the permeability coefficient for carbon dioxide determined in the temperature range from 294 to 300 K (21 to 27°C) passed through a minimum with varying degrees on the input side with increasing CO_2 pressure (up to 40 bar). Attempts were made to correlate the pressure belonging to the minimum and called plasticization pressure with the chemical structure or the physical properties of the polymers. However, a correlation could not be established between the plasticization pressure, on the one hand, and the glass transition temperature, the free volume fraction or the density of the carbonyl or sulfonyl groups in the polymers, on the other hand. Instead, it was observed that the plastification pressure occurred at a carbon dioxide concentration in the polymer of 36 ± 7 cm^3 (STP)/cm^3 and that the relation was almost linear between the ratio of permeability coefficient at plasticization pressure/permeability coefficient at a 0 bar pressure (determined by extrapolation) and the plasticization pressure. Moreover, a linear increase of the logarithm of the permeability coefficient was established when the free volume fraction increased, corresponding also to the results of other authors. However, the polyimides 8) and 9) and the polyetherimide 3) are not located on the general straight line, but form their own straight line. The results obtained in Table 49 are illustrated by the Figures 79 and 80.

Course. No.	Polymer (Trade name[1])	Fractional free volume	CO_2 plasticization pressure bar	Permeability coefficient at plasticization pressure $10^{-10} \times cm^3$(STP) cm/ cm^2 s cm Hg	Permeability coefficient at 0 bar[2] $10^{-10} \times cm^3$(STP) cm/ cm^2 s cm Hg	Measurement temperature K (°C)
1	Polysulfone (Udel® P3500)	0.152	34	3.6	5.0	296 (23)
2	Polyethersulfone (Ultrason® E 6010P)	0.152	27	2.6	3.7	294 (21)
3	Polyetherimide (Ultem® 1000)	0.180	28	0.84	1.1	294 (21)
4	Polycarbonate (Bisphenol A) (Macrolon® 3200)	0.158	31	4.7	7.5	298 (25)
5	Bisphenol-Z-poly-carbonate	0.138	24	1.0	1.4	296 (23)
6	Tetramethyl bisphenol-A polycarbonate	0.180	13	13	16	298 (25)
7	Poly-2,6-dimethyl-p-phenylene oxide	0.190	14	82	99	298 (25)
8	Polyimide (Matrimid® 5218)	0.225	12	4.8	5.7	295 (22)
9	Copolyimide (P84)	0.203	22	0.92	1.1	296 (23)
10	Cellulose acetate	0.166	11	6.0	6.7	300 (27)
11	Cellulose triacetate	0.166	10	7.3	9	297 (24)

[1] Trade names nonbinding
[2] Determined by extrapolation of the permeability coefficient/pressure curve

Table 49: Free volume fraction, CO_2 plasticization pressure and CO_2 permeability coefficient of amorphous polymers [25]

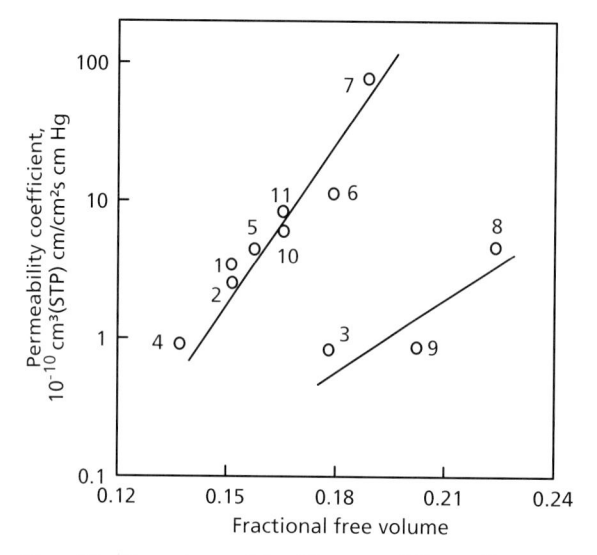

Figure 79: Dependence of the CO_2 permeability coefficient at CO_2 plastification pressure on the free volume fraction of amorphous polymers [25]
(For the number of polymers refer to Table 49)

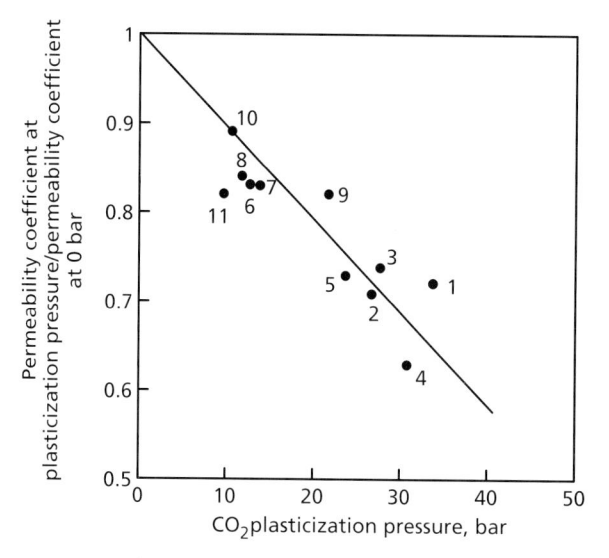

Figure 80: Relationship between the CO_2 plasticization pressure and the ratio of the CO_2 permeability coefficient at plasticization pressure and at 0 bar [25]
(For the number of polymers refer to Table 49)

Before synthetic polymers can be used for medical applications, the polymer item needs to be sterilized; apart from the established sterilization methods the treatment with liquid carbon dioxide seems to be of interest nowadays, also because it can be performed at ambient temperature. With this background, 14 commercial polymers were investigated to determine the influence of a storage of one hour in liquid carbon dioxide at 290 to 297 K (17 to 24°C) and a pressure of 6.5 MPa on the mechanical properties and the swelling behavior. In addition to pure carbon dioxide, investigations were also performed with liquid carbon dioxide to which small volumes of an aqueous H_2O_2 solution were added. It was found that the behavior of semi-crystalline polymers was more favorable than that of amorphous polymers. A statistically significant change of the tensile strength was observed for polymethylmethacrylate (PMMA), acrylonitrile-butadiene-styrene copolymer (ABS), polycarbonate (PC), polyvinyl chloride (PVC) and polyethylene terephthalate (PET) and of the modulus of elasticity in tension for polystyrene (PS) and PVC. The mass increase caused by carbon dioxide was clearly higher in case of amorphous polymers compared to the semi-crystalline polymers; the highest values were found for natural rubber (NR) elastomer (7.3%), PMMA (6.4%), polyurethane (4.5%), ABS (4.0%), ethylene-propylene-diene (EPDM) elastomer (4.0%) PS (1.9%) and PC (1.0%). A number of polymers were indicative of an extraction of oligomers, plasticizers or other components by carbon dioxide, e.g. polyurethane (PU), polyethylene (PE), PET, polyphenylene oxide (PPE) or polyvinylidene fluoride (PVDF). To assess the applicability of the new sterilization method additional investigations are considered to be necessary, in particular regarding the repeated carbon dioxide treatment of the polymers [203].

The interaction of various amorphous polymers with the supercritical carbon dioxide was examined in [204]. To this end, the thermoplastics specimens were stored in supercritical carbon dioxide at 313 or 343 K (40 or 70°C) for 1 h, followed by relaxation within 1 h. As can be seen from the mass changes of the specimens indicated in Table 50 determined after relaxation, the mass change was especially high in case of polymethylmethacrylate (PMMA), acrylonitrile-butadiene-styrene copolymer (ABS) and cellulose acetobutyrate (CAB); the specimens of these thermoplastics treated at 343 K (70°C) continued to foam up strongly during relaxation. In addition, mechanical properties of some of these thermoplastics were determined in the tensile test, revealing a clear change of mechanical properties. Prior to these tests and after having been subjected to the CO_2 treatment, the specimens had been stored in air for 30 d.

Polycarbonate films containing 2.5 to 10% of a polystyrene-polybutadiene-polystyrene triblock copolymer (SBS) were used to establish whether the treatment with supercritical carbon dioxide influences the microstructure of the polymer mixture. To this end 20 µm thick films were treated with supercritical carbon dioxide at 323 K (50°C) and a pressure of 20 MPa for 2 h. For all specimens a reduction of the size of the disperged SBS particles and a narrower particle size distribution was found compared to those without carbon dioxide treatment [205].

134 Carbon dioxide

Polymer	Thickness of specimens mm	Carbon dioxide Pressure/ temperature	Mass change Ma%	Mechanical properties in tensile test		
				Yield strength MPa	Ultimate elongation %	Elasticity modulus MPa
Polyvinyl chloride (PVC-U)	2.24	without	–	61.0	9	3006.2
		138 bar/313 K (40°C)	0.83	59.3	10	2793.1
		207 bar/313 K (40°C)	1.55	–	–	–
		207 bar/343 K (70°C)	–	49.6	12	2211.0
Polysulfone (PSU)	1.46	without	–	–	–	–
		138 bar/313 K (40°C)	2.27	–	–	–
		207 bar/313 K (40°C)	2.23	–	–	–
		207 bar/343 K (70°C)	4.09	–	–	–
Polyetherimide (PEI)	1.74	without	–	–	–	–
		138 bar/313 K (40°C)	0.88	–	–	–
		207 bar/313 K (40°C)	1.04	–	–	–
		207 bar/343 K (70°C)	2.06	–	–	–
Polystyrene, shock-resistant (PS-HI)	1.5	without	–	–	–	–
		138 bar/313 K (40°C)	2.04	–	–	–
		207 bar/313 K (40°C)	4.48	–	–	–
Acrylonitrile-butadiene-styrene copolymer (ABS)	2.40	without	–	41.9	8	1920.7
		138 bar/313 K (40°C)	4.14	43.2	6	1837.2
		207 bar/313 K (40°C)	7.24	–	–	–
		207 bar/343 K (70°C)	–	–	–	foamed up
Polyethylene terephthalate (PET), modified	2.40	without	–	–	–	–
		138 bar/313 K (40°C)	2.35	–	–	–
		207 bar/313 K (40°C)	4.18	–	–	–
Polycarbonate (PC)	3.00	without	–	67.8	69	1943.4
		138 bar/313 K (40°C)	0.93	64.9	9	2153.8
		207 bar/313 K (40°C)	1.88	–	–	–
		207 bar/343 K (70°C)	–	63.9	8	2057.2
Polymethylmethacrylate (PMMA)	3.00	without	–	63.9	3	2648.9
		138 bar/313 K (40°C)	7.86	67.2	5	2168.9
		207 bar/313 K (40°C)	11.27	–	–	–
		207 bar/343 K (70°C)	–	–	–	foamed up
Polyphenylene oxide (PPE)	6.63	without	–	–	–	–
		138 bar/313 K (40°C)	1.33	–	–	–
		207 bar/313 K (40°C)	1.32	–	–	–
		207 bar/343 K (70°C)	3.43	–	–	–
Cellulose acetobutyrate (CAB)	2.40	without	–	38.2	59	1442.7
		138 bar/313 K (40°C)	0.19	–	–	foamed up
		207 bar/313 K (40°C)	7.92	–	–	–
		207 bar/343 K (70°C)	–	–	–	foamed up

Duration of CO_2 exposure: 1 h; Duration of relaxation: 1 h
Determination of the mechanical properties after having been stored in air for 30 days

Table 50: Effects of supercritical carbon dioxide on amorphous thermoplastics [204]

Thermoplastic elastomers

Thermoplastic elastomer (TPE)	Medium	Concentration	Temperature K (°C)	Resistance	References
Polyester-TPE	CO_2 gas	technically pure	295–296 (22–23)	+	[173, 206]
Polystyrene-TPE	CO_2 gas	technically pure	RT	+	[173]
Polyolefine-TPE	CO_2 gas	technically pure	297 (24)	⊕	[173]
Polyurethane-TPE	CO_2 gas	technically pure	RT	+	[207]

+ = resistant   ⊕ = moderately resistant
RT = room temperature

Table 51: Resistance of thermoplastic elastomers to carbon dioxide gas

In [208] the permeability for carbon dioxide of thermoplastic polyurethane elastomers (TPE-U) was examined with regard to its dependence on the chemical composition in a temperature range from 288 to 333 K (15 to 60°C) using 350 μm thick foils. Here, the chemical structure of the soft segment component, the molar mass of the soft segment (polytetramethylene oxide) and the chemical structure of the diisocyanate and the chain extender, from which the hard segments were formed, were varied. The chemical composition of the TPE-U and the permeability coefficients are compiled in Table 52. As expected, the permeability coefficients strongly increased with rising temperatures. With a specified hard segment of 4,4'-diphenylmethane diisocyanate and butanediol-1,4 the lowest permeability coefficients were observed with polycaprolactone or polybutylene adipate as soft segment; the introduction of polydimethylsiloxane caused a strong increase of the permeability. The molar mass of the soft segment (polytetramethylene oxide) increasing, the permeability coefficients clearly increased up to a mean molar mass M_n of 2000 g/mol, whereas only a minor increase was observed at a molar mass above that value. With the specified soft segment the hard segments of 4,4'-diphenylmethane diisocyanate and hexanediol-1,6 exhibited the lowest permeability for carbon dioxide.

Chemical composition TPU			CO_2 permeability coefficient $10^{-13} \times cm^3$ (STP) cm /cm^2 s Pa			
Soft segment	Diisocyanate	Chain extender	288 K (15°C)	298 K (25°C)	313 K (40°C)	333 K (60°C)
a) Variation of the chemical structure of the soft segment						
Polytetramethylene oxide	4,4'-diphenyl-methane diisocyanate	butanediol-1,4	6.37	8.40	16.5	26.6

Table 52: Dependence of the permeability coefficient for carbon dioxide on the chemical structure of thermoplastic polyurethane elastomers (TPE-U) [208]

Table 52: Continued

Chemical composition TPU			CO_2 permeability coefficient $10^{-13} \times cm^3$ (STP) cm /cm² s Pa			
Soft segment	Diisocyanate	Chain extender	288 K (15°C)	298 K (25°C)	313 K (40°C)	333 K (60°C)
Polycaprolactone	4,4'-diphenyl-methane diisocyanate	butanediol-1,4	2.35	4.18	8.33	16.3
Polyhexamethylene carbonate	4,4'-diphenyl-methane diisocyanate	butanediol-1,4	1.31	2.06	5.47	12.1
Polybutylene adipate	4,4'-diphenyl-methane diisocyanate	butanediol-1,4	1.51	2.23	5.65	11.6
Block copolymer of polytetramethylene oxide and polydimethylsiloxane	4,4'-diphenyl-methane diisocyanate	butanediol-1,4	59.3	68.6	107	161
b) Variation of the molar mass of the soft segment						
Polytetramethylene oxide Mean molar mass M_n: 250 g/mol 650 g/mol 1000 g/mol 2000 g/mol 3000 g/mol	4,4'-diphenyl-methane diisocyanate	butanediol-1,4		0.083 2.00 6.24 20.1 21.8		
c) Variation of the chemical structure of the hard segment						
Polytetramethylene oxide (M_n = 2000 g/mol)	hexamethylene diisocyanate	butanediol-1,4		47.9		
Polytetramethylene oxide (M_n = 2000 g/mol)	m-xylylene diisocyanate	butanediol-1,4		25.7		
Polytetramethylene oxide (M_n = 2000 g/mol)	4,4'-diphenyl-methane diisocyanate	propanediol-1,3		16.6		

Table 52: Dependence of the permeability coefficient for carbon dioxide on the chemical structure of thermoplastic polyurethane elastomers (TPE-U) [208]

Table 52: Continued

Chemical composition TPU			CO_2 permeability coefficient $10^{-13} \times cm^3$ (STP) cm /cm^2 s Pa			
Soft segment	Diisocyanate	Chain extender	288 K (15°C)	298 K (25°C)	313 K (40°C)	333 K (60°C)
Polytetramethylene oxide (M_n = 2000 g/mol)	4,4'-diphenyl-methane diisocyanate	butanediol-1,4		19.6		
Polytetramethylene oxide (M_n = 2000 g/mol)	4,4'-diphenyl-methane diisocyanate	hexanediol-1,6		11.7		

Table 52: Dependence of the permeability coefficient for carbon dioxide on the chemical structure of thermoplastic polyurethane elastomers (TPE-U) [208]

Duroplastics

Reaction resin	Medium	Concentration	Maximum application temperature K (°C)	Resistance	References
Unsaturated polyester resins on the basis of isophthalic acid	aqueous CO_2 solution	saturated	327–350 (54–77)	+	[173]
	CO_2 gas, moist	n. d.	339–394 (66–121)	+	[173]
Unsaturated polyester resins on the basis of terephthalic acid	CO_2 gas, dry	technically pure	350 (77)	+	[173]
Unsaturated polyester resins on the basis of alcoxylated bisphenol A	aqueous CO_2 solution	saturated	305–316 (32–43)	+	[173]
	CO_2 gas, moist	n. d.	394 (121)	+	[173]
	CO_2 gas, dry	technically pure	366–450 (93–177)	+	[173]

+ = resistant ⊕ = moderately resistant
RT = room temperature
n. d. = not defined
HET acid = hexachloro-endomethylene-tetrahydrophthalic acid

Table 53: Resistance of duroplastics to aqueous carbon dioxide solutions and carbon dioxide gas

Table 53: Continued

Reaction resin	Medium	Concentration	Maximum application temperature K (°C)	Resistance	References
Unsaturated polyester resins based on HET acid	CO_2 gas, moist	n. d.	366–394 (93–121)	+	[173]
	CO_2 gas, dry	technically pure	394 (121)	+	[173]
Vinyl ester resins on the basis of bisphenol A	aqueous CO_2 solution	saturated	305 (32)	+	[173]
	CO_2 gas, moist	n. d.	372 (99)	+	[173]
	CO_2 gas, dry	technically pure	372 (99)	+	[173]
Vinyl ester resins on the basis of novolak	aqueous CO_2 solution	saturated	316 (43)	+	[173]
	CO_2 gas, moist	n. d.	366–394 (93–121)	+	[173]
	CO_2 gas, dry	technically pure	372 (99)	+	[173]
Polydial-lylphthalate	CO_2 gas	technically pure	RT	⊕	[173]
Phenolic resin	CO_2 gas	technically pure	477 (204)	+	[173]

+ = resistant   ⊕ = moderately resistant
RT = room temperature
n. d. = not defined
HET acid = hexachloro-endomethylene-tetrahydrophthalic acid

Table 53: Resistance of duroplastics to aqueous carbon dioxide solutions and carbon dioxide gas

Additional resistance data regarding duroplastics are included in section D 1 under "Organic Coatings" and D 3 "Glass fiber reinforced plastics".

Elastomers

Elastomer Rubber basis	Medium	Concentration	Temperature K (°C)	Resistance	References
Natural rubber (NR)	aqueous CO$_2$ solution	n. d.	RT	+	[158, 173, 209–212]
		n. d.	RT	⊕	[213]
	CO$_2$ gas, moist	n. d.	RT	+ to ⊕	[173, 209, 211, 212]
		n. d.	RT	⊕	[213]
	CO$_2$ gas, dry	technically pure	RT	+	[214]
		technically pure	RT	+ to ⊕	[158, 173, 209–212]
		technically pure	RT	⊕	[213]
		technically pure	333 (60)	+	[145]
Styrene butadiene rubber (SBR)	aqueous CO$_2$ solution	n. d.	RT	+	[209]
		n. d.	RT	+ to ⊕	[158, 173, 210–212]
		n. d.	RT	⊕	[213]
	CO$_2$ gas, moist	n. d.	RT	+ to ⊕	[173, 209, 211, 212]
			RT	⊕	[213]
	CO$_2$ gas, dry	technically pure	RT	+ to ⊕	[158, 173, 209–212]
		technically pure	RT	⊕	[213]
		technically pure	333 (60)	+	[145]
Chloroprene rubber (CR)	aqueous CO$_2$ solution	saturated	RT	+	[166, 167]
		saturated	≤ 353 (≤ 80)	+	[146]
		n. d.	RT	+	[150, 178, 209–212, 214]
		n. d.	RT	+ to ⊕	[158]
		n. d.	RT	⊕	[213]
	CO$_2$ gas, moist	n. d.	RT	+	[215]
		n. d.	RT	+ to ⊕	[173, 209, 211, 212]
		n. d.	RT	⊕	[213]
		n. d.	293–333 (20–60)	+	[153]
		n. d.	≤ 344 (≤ 71)	+	[166, 167, 178]
		n. d.	≤ 353 (≤ 80)	+	[146]
	CO$_2$ gas, dry	technically pure	RT	+	[167, 215]
		technically pure	RT	+ to ⊕	[158, 173, 209–212, 214]
		technically pure	RT	⊕	[213]
		technically pure	293–333 (20–60)	+	[145, 150, 153, 156]
		technically pure	≤ 344 (≤ 71)	+	[166]
		technically pure	≤ 353 (≤ 80)	+	[146]

+ = resistant   ⊕ = moderately resistant   – = not resistant
RT = room temperature
n. d. = not defined

Table 54: Resistance of elastomers to aqueous carbon dioxide solutions and carbon dioxide gas

Table 54: Continued

Elastomer Rubber basis	Medium	Concentration	Temperature K (°C)	Resistance	References
Acrylonitrile butadiene rubber (Nitrile caoutchouc) (NBR)	aqueous CO_2 solution	saturated	293–333 (20–60)	+	[163]
		saturated	≤ 353–355 (≤ 80–82)	+	[146, 166, 167]
		saturated	353 (80)	+ to ⊕	[163, 177]
		n. d.	RT	+	[150, 209, 213]
		n. d.	RT	+ to ⊕	[158, 173, 210, 211, 212, 214]
		n. d.	298–333 (25–60)	+	[152]
		n. d.	≤ 355 (≤ 82)	+	[178]
	CO_2 gas, moist	n. d.	RT	+	[173, 209, 211, 212, 213]
		n. d.	293–333 (20–60)	+	[153, 163]
		n. d.	≤ 353–355 (≤ 80–82)	+	[146, 166, 167, 178]
		n. d.	353 (80)	+ to ⊕	[163, 177]
	CO_2 gas, dry	technically pure	RT	+	[158–160, 173, 209–212, 214]
		technically pure	293–333 (20–60)	+	[145, 150, 152, 153, 156]
		technically pure	293–353 (20–80)	+	[146, 163, 166, 167]
		technically pure	353 (80)	+ to ⊕	[177]
		technically pure	373 (100)	+	[216]
Hydrogenated nitrile rubber (HNBR)	aqueous CO_2 solution	n. d.	RT	+	[209, 210]
		n. d.	RT	+ to ⊕	[211]
	CO_2 gas, moist	n. d.	RT	+	[209, 211]
	CO_2 gas, dry	technically pure	RT	+	[209–211]
Isobutene isoprene rubber (Butyl rubber) (IIR)	aqueous CO_2 solution	saturated	≤ 353 (≤ 80)	+	[146]
		n. d.	RT	+	[158, 210–212]
		n. d.	RT	+ to ⊕	[173]
		n. d.	RT	⊕	[213]
	CO_2 gas, moist	n. d.	RT	+ to ⊕	[211, 212]
		n. d.	RT	⊕	[213]
		n. d.	RT	–	[173]
		n. d.	≤ 353 (≤ 80)	+	[146]
	CO_2 gas, dry	technically pure	RT	+	[214]
		technically pure	RT	+ to ⊕	[158, 173, 210–212]
		technically pure	RT	⊕	[213]
		technically pure	333 (60)	+	[145]
		technically pure	≤ 353 (≤ 80)	+	[146]

+ = resistant   ⊕ = moderately resistant   – = not resistant
RT = room temperature
n. d. = not defined

Table 54: Resistance of elastomers to aqueous carbon dioxide solutions and carbon dioxide gas

Table 54: Continued

Elastomer Rubber basis	Medium	Concentration	Temperature K (°C)	Resistance	References
Chlorobutyl rubber (CIIR)	CO_2 gas	technically pure	RT	+	[160]
Ethylene propylene diene rubber (EPDM)	aqueous CO_2 solution	saturated	293–353 (20–80)	+	[146, 163, 177]
		saturated	≤ 372 (≤ 99)	+	[166, 167]
		n. d.	RT	+	[152, 158, 173, 209–213]
		n. d.	293–353 (20–80)	+	[150]
		n. d.	≤ 372 (≤ 99)	+	[178]
	CO_2 gas, moist	n. d.	RT	+ to ⊕	[173, 209, 211, 212]
		n. d.	RT	⊕	[213]
		n. d.	293–353 (20–80)	+	[146, 153, 163, 177]
		n. d.	≤ 366 (≤ 93)	+	[178]
		n. d.	≤ 372 (≤ 99)	+	[166, 167]
	CO_2 gas, dry	technically pure	RT	+	[152, 159, 160, 214]
		technically pure	RT	+ to ⊕	[158, 173, 209–212]
		technically pure	RT	⊕	[213]
		technically pure	333 (60)	+	[145]
		technically pure	293–353 (20–80)	+	[146, 150, 153, 156, 163, 177]
		technically pure	≤ 366 (≤ 93)	+	[166, 167]
		technically pure	373 (100)	+ to ⊕	[216]
Chlorosulfonated polyethylene (CSM)	aqueous CO_2 solution	saturated	RT	+	[166, 167]
		saturated	≤ 353 (≤ 80)	+	[146]
		n. d.	RT	+	[150, 158, 178, 209–212]
	CO_2 gas, moist	n. d.	RT	+	[209, 215]
		n. d.	RT	+ to ⊕	[211, 212]
		n. d.	293–353 (20–80)	+	[146, 153]
		n. d.	≤ 366 (≤ 93)	+	[166, 167, 178]
	CO_2 gas, dry	technically pure	RT	+	[159, 160, 214, 215]
		technically pure	RT	+ to ⊕	[158, 209–212]
		technically pure	333 (60)	+	[145]
		technically pure	293–353 (20–80)	+	[146, 150, 153, 156]
		technically pure	≤ 366 (≤ 93)	+	[166, 167]

+ = resistant   ⊕ = moderately resistant   – = not resistant
RT = room temperature
n. d. = not defined

Table 54: Resistance of elastomers to aqueous carbon dioxide solutions and carbon dioxide gas

Table 54: Continued

Elastomer Rubber basis	Medium	Concentration	Temperature K (°C)	Resistance	References
Acrylate rubber (ACM)	aqueous CO_2 solution	n. d.	RT	+	[210, 212]
		n. d.	RT	⊕	[209, 211]
		n. d.	RT	−	[213]
	CO_2 gas, moist	n. d.	RT	+ to ⊕	[211]
		n. d.	RT	⊕	[209]
		n. d.	RT	−	[213]
	CO_2 gas, dry	technically pure	RT	+ to ⊕	[211]
		technically pure	RT	⊕	[213]
		technically pure	333 (60)	+	[145]
Fluorinated rubber (FKM)	aqueous CO_2 solution	saturated	293–353 (20–80)	+	[163, 177]
		saturated	≤ 366 (≤ 93)	+	[166, 167]
		saturated	373 (100)	+	[146]
		saturated	373 (100)	+ to ⊕	[163]
		n. d.	RT	+	[173, 209–214]
		n. d.	RT	+ to ⊕	[158]
		n. d.	298–333 (25–60)	+	[152]
		n. d.	293–353 (20–80)	+	[150]
		n. d.	≤ 366 (≤ 93)	+	[178]
	CO_2 gas, moist	n. d.	RT	+	[209, 213, 215]
		n. d.	RT	+ to ⊕	[173, 211, 212]
		n. d.	293–353 (20–80)	+	[153, 177]
		n. d.	≤ 366 (≤ 93)	+	[166, 167, 178]
		n. d.	293–393 (20–120)	+	[146, 163]
	CO_2 gas, dry	technically pure	RT	+	[152, 213, 215]
		technically pure	RT	+ to ⊕	[158, 173, 209–212, 214]
		technically pure	298–333 (25–60)	+	[145]
		technically pure	293–353 (20–80)	+	[150, 153, 177]
		technically pure	≤ 366 (≤ 93)	+	[166, 167]
		technically pure	293–373 (20–100)	+	[156, 216]
		technically pure	293–393 (20–120)	+	[146, 163]

+ = resistant   ⊕ = moderately resistant   − = not resistant
RT = room temperature
n. d. = not defined

Table 54: Resistance of elastomers to aqueous carbon dioxide solutions and carbon dioxide gas

Table 54: Continued

Elastomer Rubber basis	Medium	Concentration	Temperature K (°C)	Resistance	References
Perfluoro rubber (FFKM)	aqueous CO$_2$ solution	any n. d.	373 (100) RT	+ +	[173] [209–211]
	CO$_2$ gas, moist	n. d.	RT	+	[209, 211]
	CO$_2$ gas, dry	any technically pure	373 (100) RT	+ +	[173, 216] [209–211]
Silicone rubber (MQ, VMQ)	aqueous CO$_2$ solution	n. d. n. d.	RT RT	+ ⊕	[173, 209–212] [213]
	CO$_2$ gas, moist	n. d. n. d. n. d.	RT RT RT	+ + to ⊕ ⊕	[209, 215] [173, 211, 212] [213]
	CO$_2$ gas, dry	technically pure technically pure technically pure technically pure technically pure	RT RT RT 333 (60) 373 (100)	+ + to ⊕ ⊕ + + to ⊕	[214, 215] [173, 209–212] [213] [145] [216]
Fluorosilicone rubber (FVMQ)	aqueous CO$_2$ solution	n. d. n. d.	RT RT	+ ⊕	[173, 209–212] [213]
	CO$_2$ gas, moist	n. d. n. d. n. d.	RT RT RT	+ + to ⊕ ⊕	[209] [173, 211, 212] [213]
	CO$_2$ gas, dry	technically pure technically pure technical grade technical grade technically pure	RT RT RT 333 (60) 373 (100)	+ + to ⊕ ⊕ + +	[210] [173, 209, 211, 212] [213] [145] [216]

+ = resistant ⊕ = moderately resistant – = not resistant
RT = room temperature
n. d. = not defined

Table 54: Resistance of elastomers to aqueous carbon dioxide solutions and carbon dioxide gas

Table 54: Continued

Elastomer Rubber basis	Medium	Concentration	Temperature K (°C)	Resistance	References
Polyurethane rubber (AU, EU)	aqueous CO_2 solution	n. d.	RT	+	[158, 173, 177, 210, 212, 214, 217]
		n. d.	RT	⊕	[209, 211, 213]
	CO_2 gas, moist	n. d.	RT	+	[177, 215]
		n. d.	RT	⊕	[209, 213]
	CO_2 gas, dry	technically pure	RT	+	[158, 173, 177, 209–212, 214, 215]
		technically pure	RT	⊕	[213]
		technically pure	333 (60)	–	[145]
Polysulphide rubber (OT)	aqueous CO_2 solution	n. d.	RT	+	[173, 212]
	CO_2 gas, moist	n. d.	RT	+	[210]
		n. d.	RT	+ to ⊕	[173, 212]
	CO_2 gas, dry	technically pure	RT	+ to ⊕	[173, 210, 212]
		technically pure	373 (100)	+ to ⊕	[216]
Epichlorhydrin rubber (ECO)	aqueous CO_2 solution	n. d.	RT	+	[210, 211]
	CO_2 gas, moist	n. d.	RT	+	[211]
	CO_2 gas, dry	technically pure	RT	+	[210–212]

+ = resistant ⊕ = moderately resistant – = not resistant
RT = room temperature
n. d. = not defined

Table 54: Resistance of elastomers to aqueous carbon dioxide solutions and carbon dioxide gas

Most of the elastomers are resistant to aqueous carbon dioxide solutions or carbon dioxide gas (moist or dry) at ambient conditions; for several elastomers an upper application temperature of 333 to 353 K (60 to 80°C) (NBR, CSM) or 353 to 373 K (80 to 100°C) (EPDM, FKM) is indicated. The highest resistance and temperature of use are exhibited by elastomers on the basis of FFKM.

The resistance figures in Table 54 for carbon dioxide gas mainly refer to low and moderate CO_2 pressures. At high CO_2 pressures an increasing sorption of the carbon dioxide in the polymer needs to be taken into account. One FKM elastomer exhibited a particularly critical behavior as revealed by the investigations in [218]. The volume increase of FKM elastomer specimens caused by sorption was almost linear at 308 K (35°C) as the CO_2 pressure increased to 130 bar and reached a value of 50 % at 120 bar. Irrespective of this marked deformation a sudden pressure drop caused severe damage to the material due to bubble formation. Therefore, it is not recommended to use FKM elastomers for sealing high-pressure vessels or use them as sealants in pumps, valves and similar equipment. Among the examined materials, one EPDM elastomer exhibited the highest resistance to highly compressed carbon dioxide.

Also in [39] reference is made to the risk of damage to the material by a sudden pressure drop, which is also called explosive decompression. With reference to its use as a sealant in CO_2 air conditioning systems where operating pressures of CO_2 exceeding 100 bar and temperature variations from 233 to 473 K (–40°C to 200°C) may occur, the permeability of elastomers on the basis of FKM, NBR, HNBR and EPDM as well as of several thermoplastic materials was investigated with reference to its dependence on pressure in a temperature range from 293 to 393 K (20°C to 120°C). As can be seen from Figure 81, the permeability coefficient for carbon dioxide of an FKM elastomer measured at 293 K (20°C) strongly increased at CO_2 pressures above 30 bar and elastomers on the basis of EPDM or HNBR exhibited a much lower dependence of the permeability coefficient. The steep increase of the permeability coefficient of the FKM elastomer observed at 293 K (20°C) became shallower when the test temperature increased, as shown in Figure 82. The example of two HNBR qualities of different origin was used to demonstrate that the CO_2 permeability coefficients may strongly depend on the chemical composition; on average, the permeability coefficients of the HNBR elastomer I were about 50 % above those of the HNBR elastomer II (refer to Figure 83).

In [219] two silicone elastomers, i.e. an FKM elastomer and an NBR elastomer saturated with carbon dioxide at 40 bar, and the changes of the mechanical properties in

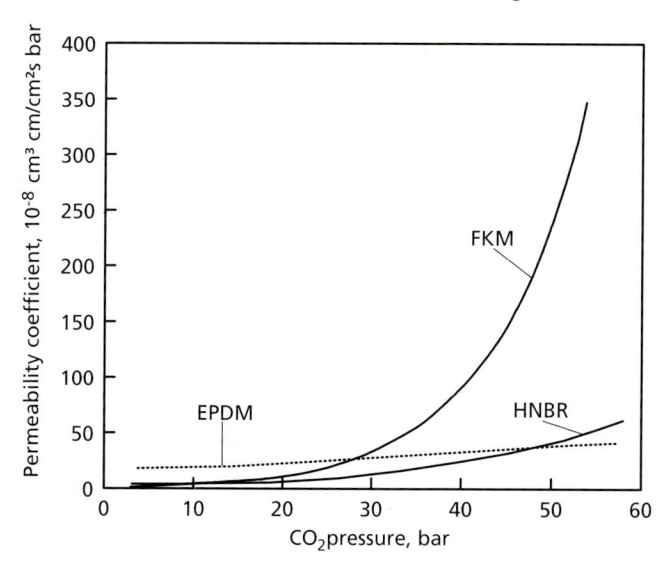

Figure 81: Dependence of the CO_2 permeability coefficient of several elastomers on the CO_2 pressure at 293 K (20°C) [39]

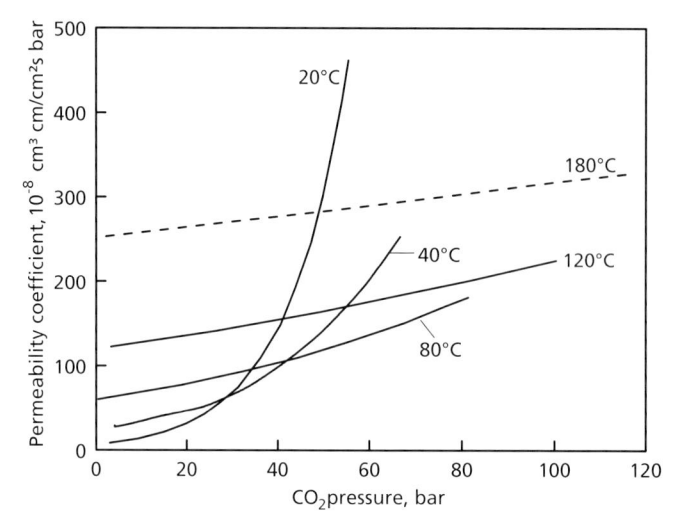

Figure 82: Dependence of a CO_2 permeability coefficient of an FKM elastomer on the CO_2 pressure at various temperatures [39]

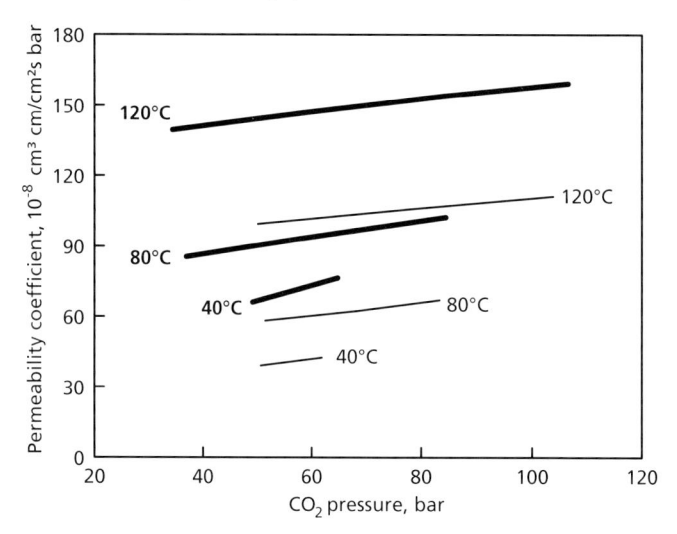

Figure 83: Dependence of the CO_2 permeability coefficients of two HNBR elastomers of different origin on the CO_2 pressure at different temperatures [39]
— HNBR I　　—— HNBR II

the temperature range from 293 to 453 K (20 to 180°C) were examined. As the data in Table 55 shows, the ultimate tensile strength, the elongation at break and the tear strength, both in air and following saturation with carbon dioxide decreased as the temperature increased and that the drop of the strength values caused by sorption of the carbon dioxide was more marked in the lower temperature range compared to high temperatures. In addition to the strength tests, the silicone elastomer specimens were saturated with carbon dioxide at a pressure of 40 bar and at a temperature of 293 to 298 K (20 to 25°C) and then heated to a temperature of 423 K (150°C) at a heating rate of 60°C/min. The silicone elastomer A survived 10 cycles unscathed, whereas the silicone elastomer B exhibited strong decomposition after 5 cycles. The NBR elastomer and the FKM elastomer heated to 423 K (150°C) at a heating rate of 5°C/min exhibited a strong crack formation after 10 cycles. Moreover, the silicone elastomer A and the NBR elastomer were subjected a repeated reduction of the CO_2 pressure from 40 to 3.5 bar at a rate of 10 bar/min at 293, 323 and 373 K (20, 50 and 100°C); after 10 cycles detrimental changes had not occurred.

Temperature K (°C)	Ultimate tensile strength		Elongation at break		Tear strength	
	Without stress N/mm^2	Preservation after CO_2 saturation %	Without stress %	Preservation after CO_2 saturation %	Without stress N/mm^2	Preservation after CO_2 saturation %
a) NBR elastomer						
293 (20)	15	80	260	81	21	38
323 (50)	14	86	240	92	11	55
353 (80)	12	92	220	86	10	60
383 (110)	11	91	200	90	4	150
b) FKM elastomer						
293 (20)	14	43	200	60	20	15
323 (50)	11	54	180	67	5	40
353 (80)	8	63	130	77	3	67
383 (110)	6	67	110	73	2	50
423 (150)	5	100	90	89	1	100
453 (180)	5	100	70	100	1	100
c) Silicone elastomer A						
293 (20)	7.5	46	360	56	4.8	38
323 (50)	5.0	70	320	75	2.8	61
353 (80)	4.5	78	280	79	1.5	106
383 (110)	4.5	89	220	95	1.3	100
453 (180)	3.0	100	180	94	1.0	100
d) Silicone elastomer B						
293 (20)	6.0	67	170	76	2.0	50

Table 55: Change in the mechanical properties of elastomers following saturation with carbon dioxide at 40 bar [219]

Additional resistance specifications are included in Section D 2 under "O-rings".

D
Materials with special properties

D 1   Coatings and linings

Organic coatings and linings

For decades steel and concrete components in process plants, e.g. plants of the chemical industry, have been widely protected against corrosion by means of organic materials in the form of coatings and linings (rubber linings and thermoplastic linings).

Thin coatings for concrete

Coatings on the basis of aqueous polymer dispersions are widely used to protect concrete surfaces against atmospheric influences. In addition to other requirements, these coatings must have a lowest possible permeability coefficient for carbon dioxide to prevent the carbonatization of concrete and exhibit a high water vapor permeability to facilitate the exchange of moisture with the environment. This requirement is met by numerous coatings offered on the market, such as coatings on the basis of pigmented aqueous polymer dispersions as proven much earlier by relevant investigations (including, without limitation, in [220]).

In [221] the influence of the chemical composition of acrylic copolymer produced by emulsion polymerization in the aqueous phase was investigated to determine the permeability for carbon dioxide and water vapor. In a single-stage emulsion polymerization process 73.7 % by weight of a main monomer were polymerized with 23.5 % by weight of a second monomer and 2.8 % by weight methacrylic acid as another comonomer. The permeation measurements were performed at 298 K (25°C) with the polymer films obtained by the evaporation of water and drying in a vacuum, the pressure difference of carbon dioxide being 1 atm, double-distilled water being used for the water vapor measurements and the relative air humidity being set to 50 % on the rear side. Table 56 shows that a low permeability coefficient for carbon dioxide and a water vapor permeability judged as good were only obtained when ethyl acrylate was used as the main monomer, in particular when acrylonitrile was used as the second monomer. Although a polymer consisting of 97.2 % by weight ethyl acrylate and 2.8 % by weight methacrylic acid used for comparative testing (without using a second monomer) exhibited a very good water vapor permeability, its CO_2 permeability was undesirably high.

Main monomer 73.7 Ma%	Second monomer 23.5 Ma%	Glass transition temperature K (°C)	Permeability coefficient for carbon dioxide at 298 K (25°C) $10^{-13} \times cm^3$(STP) cm/cm^2 s Pa	Permeability coefficient for water vapor at 298 K (25°C) $10^{-2} \times g\ cm/d\ m^2$ mm Hg
Ethyl acrylate	acrylonitrile	271.65 (−1.5)	9.90	3.9
Butyl acrylate	acrylonitrile	247 (−26.0)	66.0	4.0
2-Ethylhexyl acrylate	acrylonitrile	240.55 (−32.6)	81.5	1.7
Ethyl acrylate	methyl-methacrylate	272.25 (−0.9)	20.3	3.7
Butyl acrylate	methyl-methacrylate	247.65 (−25.5)	73.4	2.8
2-Ethylhexyl acrylate	methyl-methacrylate	240.55 (−32.6)	77.9	8.2
Ethyl acrylate	styrene	271.05 (−2.1)	21.8	3.5
Butyl acrylate	styrene	246.55 (−26.6)	65.6	2.5
2-Ethylhexyl acrylate	styrene	240.05 (−33.1)	81.1	8.8
Ethyl acrylate	without		69.5	6.1

Comonomer: 2.8 % by weight methacrylic acid

Table 56: Influence of the chemical composition of acryl copolymers on the permeability for carbon dioxide and water vapor in water-based coatings [221]

Powder coatings

Coating Chemical basis (Trade name[1])	Layer thickness mm	Medium	Water quality	Temperature K (°C)	Resistance	References
Polyethylene, low density (PE-LD), modified	0.3–0.6	aqueous CO_2 solution	n. d.	294–333 (21–60)	+	[222]
	0.3–0.6	CO_2 gas	technically pure	294–333 (21–60)	+	[222]
Polyethylene ionomer (Epover®-S)	0.25–0.5	aqueous CO_2 solution	n. d.	294–333 (21–60)	+	[223]
	0.25–0.5	CO_2 gas	technically pure	294–333 (21–60)	+	[223]
Polyolefin mixture (Plascoat® PPA 571)	0.3–0.75	aqueous CO_2 solution	n. d.	≤ 333 (≤ 60)	+	[224]

+ = resistant   ⊕ = moderately resistant
n. d. = not defined
[1] Trade names nonbinding

Table 57: Resistance of powder coatings to aqueous carbon dioxide solutions and carbon dioxide gas

Table 57: Continued

Coating Chemical basis (Trade name[1)])	Layer thickness mm	Medium	Water quality	Temperature K (°C)	Resistance	References
Polyamide 11 (PA11) (Rilsan® PA 11)	0.1–0.8	aqueous CO_2 solution	n. d.	293–363 (20–90)	+	[225]
Ethylene-chloro-trifluoroethylene copolymer (ECTFE) (E-CTFE Halar®)	0.4–1.0	aqueous CO_2 solution	n. d.	296–423 (23–150)	+	[226, 227]
	0.4–1.0	CO_2 gas, moist	n. d.	296–423 (23–150)	+	[226, 227]
	0.4–1.0	CO_2 gas, dry	technically pure	296–423 (23–150)	+	[226, 227]
Ethylene-tetrafluoro-ethylene copolymer (ETFE); (Fluon® ETFE Rotomolding)	2–5	CO_2 gas	technically pure	298–373 (25–100)	+	[228]
			technically pure	383 (110)	+ to ⊕	[228]
Tetrafluoroethylene-perfluoropropylvinyl-ether copolymer (PFA) (Edlon® PFA)	max. 1	aqueous CO_2 solution	any	≤ 403 (≤ 130)	+	[229]
	max. 1	CO_2 gas	any	≤ 403 (≤ 130)	+	[229]
ETFE Intermediate layer = high-melt ETFE about 0.7 mm, cover layer = low-melt ETFE about 0.8 mm (Rhenoguard® Jumbo I)	1.2–1.8	CO_2 gas	technically pure	298–373 (25–100)	+	[230]
			technically pure	383 (110)	⊕	[230]
PFA/tetrafluoro-ethylene-hexafluoro-propylene copolymer (FEP) Intermediate layer = PFA, cover layer = FEP (Rhenoguard® Jumbo II)	1.0–1.2	aqueous CO_2 solution	any	≤ 413 (≤ 140)	+	[230]
	1.0–1.2	CO_2 gas, moist	any	≤ 413 (≤ 140)	+	[230]
	1.0–1.2	CO_2 gas, dry	any	≤ 413 (≤ 140)	+	[230]

+ = resistant ⊕ = moderately resistant
n. d. = not defined
[1)] Trade names nonbinding

Table 57: Resistance of powder coatings to aqueous carbon dioxide solutions and carbon dioxide gas

Table 57: Continued

Coating Chemical basis (Trade name[1])	Layer thickness mm	Medium	Water quality	Temperature K (°C)	Resistance	References
PFA/FEP Intermediate layer = PFA filled 0.6 mm, cover layer = FEP filled 0.6 mm, topcoat = FEP 0.2 mm (Rhenoguard® Jumbo III)	1.2–1.8	aqueous CO_2 solution	any	≤ 433 (≤ 160)	+	[230]
	1.2–1.8	CO_2 gas, moist	any	≤ 433 (≤ 160)	+	[230]
	1.2–1.8	CO_2 gas, dry	any	≤ 433 (≤ 160)	+	[230]

+ = resistant ⊕ = moderately resistant
n. d. = not defined
[1] Trade names nonbinding

Table 57: Resistance of powder coatings to aqueous carbon dioxide solutions and carbon dioxide gas

Powder coatings on the basis of PE and PA11 can be used up to about 333 K (60°C) against aqueous carbon dioxide solutions or carbon dioxide gas. The upper application temperature of powder coatings on the basis of partially fluorated thermoplastics ECTFE and ETFE is higher (about 373 K (100°C)) and the highest resistance is reached with the fully fluorinated thermoplastics PFA and FEP (403 to 433 K (130 to 160°C)).

Reaction resin coatings for vessels

Coating Chemical basis (Trade name[1])	Layer thickness mm	Medium	Concentration	Temperature K (°C)	Resistance	References
a) Spray-applied coatings						
Phenolic resin, unmodified, thermosetting	0.2	aqueous CO_2 solution	n. d.	RT	⊕	[231]
		CO_2 gas, dry	technically pure	RT	+	[231]
Phenolic resin/EP resin, heat curing (Proco® – L)	0.2–0.25	CO_2 gas	technically pure	RT	+	[232]

+ = resistant ⊕ = moderately resistant
RT = room temperature
n. d. = not defined
VE = vinyl ester
UP = unsaturated polyester
EP = epoxide
HET acid = hexachloro-endomethylene-tetrahydrophthalic acid
[1] Trade names nonbinding

Table 58: Resistance of coatings to aqueous carbon dioxide solutions and carbon dioxide gas

Table 58: Continued

Coating Chemical basis (Trade name[1])	Layer thickness mm	Medium	Concentration	Temperature K (°C)	Resistance	References
EP resin/tar	0.3–0.4	CO_2 gas	technically pure	RT	+	[231]
Polyurethane/tar	0.3–0.4	CO_2 gas	technically pure	RT	+	[231]
VE novolak resin/ glass flakes (Chempruf® 141)	0.5–1.5	aqueous CO_2 solution	saturated	305 (32)	+	[233]
	0.5–1.5	CO_2 gas, moist	n. d.	≤ 327 (≤ 54)	+	[233]
VE novolak resin/ mineral flakes (Bücolit® V590G)	1.5	CO_2 gas	technically pure	≤ 453 (≤ 180)	+	[234]
UP-Bisphenol resol/ glass flakes (Chempruf® 131)	0.5–1.5	aqueous CO_2 solution	saturated	305 (32)	+	[233]
		CO_2 gas, moist	n. d.	≤ 327 (≤ 54)	+	[233]
UP HET acid resin/ glass flakes (Chempruf® 130)	0.5–1.5	aqueous CO_2 solution	saturated	≤ 327 (≤ 54)	+	[233]
	0.5–1.5	CO_2 gas, moist	n. d.	≤ 327 (≤ 54)	+	[233]
EP novolak resin/ polyamine/glass flakes (Chempruf® 121)	0.5–1.5	aqueous CO_2 solution	saturated	300 (27)	+	[233]
	0.5–1.5	CO_2 gas, moist	n. d.	≤ 355 (≤ 82)	+	[233]
EP-Novolak resin/ polyamine/aluminium oxide and other fillers (Proguard® CN 100)	1.0	aqueous CO_2 solution	n. d.	≤ 333 (≤ 60)	+	[235]
EP novolak resin/ polyamine/flake filler (Plasguard® 4550)	0.8–1.3	CO_2 gas	technically pure	≤ 327 (≤ 54)	+	[236]

+ = resistant
RT = room temperature
n. d. = not defined
VE = vinyl ester
UP = unsaturated polyester
EP = epoxide
HET acid = hexachloro-endomethylene-tetrahydrophthalic acid
[1] Trade names nonbinding

Table 58: Resistance of coatings to aqueous carbon dioxide solutions and carbon dioxide gas

Table 58: Continued

Coating Chemical basis (Trade name[1])	Layer thickness mm	Medium	Concentration	Temperature K (°C)	Resistance	References
EP bisphenol resol/ polyamine/glass flakes (Chempruf® 120)	0.5–1.5	aqueous CO_2 solution	saturated	300 (27)	+	[233]
	0.5–1.5	CO_2 gas, moist	n. d.	≤ 355 (≤ 82)	+	[233]
Polyurethane/fillers	0.5–2	aqueous CO_2 solution	any	RT	+	[237]
b) Trowel-applied coatings						
VE- Novolak resin/ glass flakes (Chempruf® 1410)	1.5–2.5	aqueous CO_2 solution	saturated	≤ 316 (≤ 43)	+	[233]
	1.5–2.5	CO_2 gas, moist	n. d.	≤ 355 (≤ 82)	+	[233]
VE resin/glass flakes (Oxydur® Flake)	2	CO_2 gas	technically pure	293 (20)	+	[238]
UP-bisphenol resol/ glass flakes (Chempruf® 1310)	1.5–2.5	aqueous CO_2 solution	saturated	305 (32)	+	[233]
	1.5–2.5	CO_2 gas, moist	n. d.	≤ 344 (≤ 71)	+	[233]
UP HET acid resin/ glass flakes (Chempruf® 1300)	1.5–2.5	aqueous CO_2 solution	saturated	≤ 333 (≤ 60)	+	[233]
	1.5–2.5	CO_2 gas, moist	n. d.	≤ 355 (≤ 82)	+	[233]
Furan resin/fillers	2–3	aqueous CO_2 solution	any	RT	+	[237]
Phenolic resin/fillers	2–3	aqueous CO_2 solution	any	RT	+	[237]
Polyurethane/fillers	2–3	aqueous CO_2 solution	any	RT	+	[237]
c) Laminate coatings						
VE novolak resin/ glass mats/glass fleece (Bücolit® V47-36, Chempruf® 2410)	3.2	aqueous CO_2 solution	saturated	≤ 316 (≤ 43)	+	[233]
	3.2	CO_2 gas, moist	n. d.	≤ 355 (≤ 82)	+	[233]
	2	CO_2 gas	technically pure	≤ 453 (≤ 180)	+	[234]

+ = resistant
RT = room temperature
n. d. = not defined
VE = vinyl ester
UP = unsaturated polyester
EP = epoxide
HET acid = hexachloro-endomethylene-tetrahydrophthalic acid
[1] Trade names nonbinding

Table 58: Resistance of coatings to aqueous carbon dioxide solutions and carbon dioxide gas

Table 58: Continued

Coating Chemical basis (Trade name[1])	Layer thickness mm	Medium	Concentration	Temperature K (°C)	Resistance	References
VE resin/glass mats/ glass fleece (Oxydur® VE-L)	2.5	CO_2 gas	technically pure	293 (20)	+	[238]
UP bisphenol A resin/glass mats/ glass fleece (Chempruf® 2310)	3.2	aqueous CO_2 solution	saturated	≤ 305 (≤ 32)	+	[233]
	3.2	CO_2 gas, moist	n. d.	≤ 344 (≤ 71)	+	[233]
UP HET acid resin/ glass mats/glass fleece (Chempruf® 2300)	3.2	aqueous CO_2 solution	saturated	≤ 344 (≤ 71)	+	[233]
	3.2	CO_2 gas, moist	n. d.	≤ 355 (≤ 82)	+	[233]
UP isophthalic acid resin/glass mats/ glass fleece (Bücolit® V25)	2	aqueous CO_2 solution	n. d.	RT	+	[234]
	2	CO_2 gas	technically pure	≤ 373 (≤ 100)	+	[234]
EP bisphenol A resin/polyamine/ glass mats/glass fleece (Chempruf® 2201)	3.2	aqueous CO_2 solution	saturated	≤ 355 (≤ 82)	+	[233]
	3.2	CO_2 gas, moist	n. d.	≤ 322 (≤ 49)	+	[233]
Phenolic resin/glass-fiber mats/fleece	2–4	aqueous CO_2 solution	any	RT	+	[237]
Furan resin/glass-fiber mats/fleece (Chempruf® 2101)	3.2	aqueous CO_2 solution	saturated	≤ 394 (≤ 121)	+	[233]
	3.2	CO_2 gas, moist	n. d.	≤ 377 (≤ 104)	+	[233]

+ = resistant
RT = room temperature
n. d. = not defined
VE = vinyl ester
UP = unsaturated polyester
EP = epoxide
HET acid = hexachloro-endomethylene-tetrahydrophthalic acid
[1] Trade names nonbinding

Table 58: Resistance of coatings to aqueous carbon dioxide solutions and carbon dioxide gas

For an exposure to aqueous carbon dioxide solutions spray-applied coatings on the basis of VE and UP resins can be used up to 303 to 323 K (30 to 50°C) and on the basis of EP resins up to 303 to 333 K (30 to 60°C); for the thicker trowel-applied and laminate coatings the upper application temperature is mostly 10 to 20°C higher. The resistance to moist carbon dioxide is classified a bit higher than to aqueous carbon dioxide solutions such that the upper temperature of use can amount up to 353 K (80°C). Dry carbon dioxide gas is considered to be even less critical, reflected by a maximum application temperature of 453 K (180°C) for VE novolak resin coatings.

Combination linings

Combined linings consist of ceramic or carbon boards or tiles laid and grouted in reaction resin putties on a sealing layer of reaction resin coatings, thermoplastic or elastomer webs. Ceramic and carbon tiles or bricks are resistant to aqueous carbon dioxide solutions at ambient conditions and at an elevated temperature (up to about 403 K (130°C)) [239]. The resistance of the laying or joint-filling putties and the sealing layers is indicated in Table 59.

Polymer/reaction resin (Trade name[1])	Medium	Concentration	Temperature K (°C)	Resistance	References
a) Reaction resin putties					
Furan resin/mineral fillers (Furadur®-Kitt, Keranol® FU 310)	aqueous CO_2 solution	saturated	293–413 (20–140)	+	[239]
	CO_2 gas	technically pure	RT	+	[240, 241]
Furan resin/carbon fillers (Asplit® FN, Furadur®-F-Kitt, Keranol® FU 320)	aqueous CO_2 solution	saturated	293–413 (20–140)	+	[239]
		n. d.	approx. 373 (100)	+	[242]
	CO_2 gas	technically pure	RT	+	[240, 241]
Phenolic resin/mineral fillers	aqueous CO_2 solution	any	RT	+	[243, 244]
Phenolic resin, modified/ carbon fillers (Asplit® CN)	aqueous CO_2 solution	n. d.	approx. 373 (100)	+	[242]

+ = resistant   ⊕ = moderately resistant   – = not resistant
RT = room temperature
n. d. = not defined
[1] Trade names nonbinding

Table 59: Resistance of putties and sealing layers to aqueous carbon dioxide solutions and carbon dioxide

Table 59: Continued

Polymer/reaction resin (Trade name[1])	Medium	Concentration	Temperature K (°C)	Resistance	References
Vinyl ester resins/mineral fillers (Asplit® VEQ, Keranol® VE 311, Oxydur® VE-K)	aqueous CO_2 solution	any	RT	+	[243, 244]
		saturated	293–393 (20–120)	+	[239]
		n. d.	293 (20)	+	[242]
	CO_2 gas	technically pure	293 (20)	+	[245]
		technically pure	≤ 394 (≤ 121)	+	[240]
Vinyl ester resin/carbon fillers (Asplit® VEC)	aqueous CO_2 solution	any	RT	+	[243, 244]
		n. d.	293 (20)	+	[242]
	CO_2 gas	technically pure	≤ 366 (≤ 93)	+	[240]
Unsaturated polyester resin/mineral fillers (Keranol® UP 311, Oxydur® TM)	aqueous CO_2 solution	any	RT	+	[243, 244]
		saturated	293–373 (20–100)	+	[239]
	CO_2 gas	technically pure	293 (20)	+	[245]
Epoxy resin/polyamine/mineral fillers (Asplit® ET, Keranol® EP 310, Alkadur® K 75)	aqueous CO_2 solution	saturated	293–353 (20–80)	+	[239]
		n. d.	293 (20)	+	[242]
	CO_2 gas	technically pure	RT	+	[240, 246]
Potassium silicate/mineral fillers (Asplit® HB, HES)	aqueous CO_2 solution	n. d.	RT	−	[242]
	CO_2 gas	technically pure	RT	+	[240]
Bitumen/mineral fillers	aqueous CO_2 solution	any	RT	+	[244]
Cement mortar	aqueous CO_2 solution	any	RT	−	[244]
b) Sealing layers					
b1) Coatings (self-leveling and trowel-applied coatings, thickness 1.5–3 mm)					
Epoxy resin/polyamine/fillers	aqueous CO_2 solution	any	RT	+ to ⊕	[247]
Polyurethane/fillers (Oxydur® HT, UP 82)-polyurethane	CO_2 gas	technically pure	RT	+	[240]
		technically pure	293–323 (20–50)	+	[248]

+ = resistant ⊕ = moderately resistant − = not resistant
RT = room temperature
n. d. = not defined
[1] Trade names nonbinding

Table 59: Resistance of putties and sealing layers to aqueous carbon dioxide solutions and carbon dioxide

Table 59: Continued

Polymer/reaction resin (Trade name[1])	Medium	Concentration	Temperature K (°C)	Resistance	References
Phenolic resin/fillers	aqueous CO_2 solution	any	RT	+	[247]
Furan resin/fillers	aqueous CO_2 solution	any	RT	+	[247]
Vinyl ester resin/fillers	aqueous CO_2 solution	any	RT	+	[247]
Unsaturated polyester resin/fillers	aqueous CO_2 solution	any	RT	+	[247]
b2) Elastomer or rubber coatings (thickness about 2 mm)					
Butyl rubber (IIR)	aqueous CO_2 solution	any	RT	+	[247]
Bromobutyl rubber (BIIR)	aqueous CO_2 solution	any	RT	+	[247]
Chlorobutyl rubber (CIIR)	aqueous CO_2 solution	any	RT	+	[247]
Chloroprene rubber (CR)	aqueous CO_2 solution	any	RT	+	[247]
Acrylonitrile butadiene rubber (NBR)	aqueous CO_2 solution	any	RT	+	[247]
Chlorosulfonated polyethylene (CSM)	aqueous CO_2 solution	any	RT	+	[247]
b3) Thermoplastic webs (thickness approx. 2 mm)					
Polyisobutylene (PIB)/carbon fillers (Rhepanol® O.R.G.)	aqueous CO_2 solution	n. d.	293–353 (20–80)	+	[249]
Polyisobutylene (PIB)/mineral fillers (Rhepanol® O.R.F.)	aqueous CO_2 solution	any n. d.	RT 293–353 (20–80)	+ +	[247] [249]
Polyvinyl chloride with plasticizers (PVC-P)	aqueous CO_2 solution	any	RT	+ to –	[247]

+ = resistant ⊕ = moderately resistant – = not resistant
RT = room temperature
n. d. = not defined
[1] Trade names nonbinding

Table 59: Resistance of putties and sealing layers to aqueous carbon dioxide solutions and carbon dioxide

Among the reaction resin putties, putties on the basis of furan and phenolic resin exhibit the highest resistance to aqueous carbon dioxide solutions and moist carbon dioxide gas, followed by VE and UP resin putties, whereas EP resin putties can be used to a maximum temperature of 353 K (80°C) only. Cement mortar waterglass cements are not recommended.

Suitable sealing layers are coatings of the known reaction resins, but also elastomer sheet on the basis of IIR, BIIR, CIIR, CR, NBR or CSM or thermoplastic sheets on the basis of PIB or PVC-P.

Industrial floor coatings

Floor coatings (thickness 1 to 5 mm) and stone floors (thickness > 5 mm) on the basis of the known reaction resins (epoxy resins, polyurethane, unsaturated polyester resins, vinyl ester resins) are resistant to aqueous carbon dioxide solutions (of any concentration) at ambient conditions and can often be used even at elevated temperatures (up to 333 K (60°C)) [239, 240, 248, 250–252].

Coatings for water protection

As coatings for retention ponds, tanks and areas made of concrete in facilities for the storage, filling and handling of liquids hazardous to water, which in Germany require a general construction supervisory approval of the German Institute for Construction Engineering (Deutsches Institut für Bautechnik – DIBt), are resistant in contact with aqueous carbon dioxide solutions (concentration: any) about 2 mm thick epoxy resin coatings, such as Keracid® EP 1102 [253], Sikafloor® water protection system 390 [254], MC protection system 1900 [252], Disboxid® water protection system WHG Standard [255], Alkadur® HR [256] or StoCretec WHG System 1 [257].

Rubber linings

Rubber linings are linings with rubber webs or elastomer webs that are bonded to the steel or concrete parts over their entire surface area using adhesives. The layer thickness of the rubber webs is approx. 3–6 mm; if particularly high stress is involved, two rubber webs are bonded on top of each other. The rubber webs can consist of a single layer or of several different layers, for example two layer webs consisting of a chemically resistant main layer and an easily bonded lower layer are often utilized in practice. Factory rubber linings are carried out on steel components that are transportable and do not exceed a certain maximum size, otherwise the rubber linings on the construction site are carried out on very large steel components and on concrete components.

Hard rubber linings are highly crosslinked polymer materials with duroplastic character and no rubber elastic properties at all. Mainly natural rubber (NR), isoprene rubber (IR) and styrene-butadiene rubber (SBR) are used as rubber materials for hard rubber linings. Hard rubber linings feature broad chemical durability, e.g. with respect to acids, alkalis, salt solutions and organic solvents, but their resistance to oxidizing substances is limited. Semi-hard rubber linings exhibit a lower degree of cross-linking compared to hard rubber linings, but are rather associated with hard rubber linings due to their properties than to soft rubber linings.

For soft rubber linings the preferred rubber material is butyl rubber (IIR) including the halogenated variants chlorobutyl rubber (CIIR) and bromobutyl rubber (BIIR) since these rubbers are characterized by a low permeability to water vapor, oxygen, sulfur dioxide, carbon dioxide and other compounds of low molecular weight as well as by a wide chemical resistance, e.g. to acids, alkalis and salt solutions, and a very good resistance to weathering and aging resistance. Soft rubber linings on the basis of chloroprene rubber (CR) also exhibit a good chemical resistance, even to oils and greases, as well as a high abrasion resistance. However, their permeability is significantly higher than that of the IIR/CIIR/BIIR rubber linings. Natural rubber (NR) is utilized for soft rubber linings chiefly when high wear and abrasion resistance is demanded and slight to moderate chemical stress is encountered. Soft rubber linings on the basis of chlorosulfonated polyethylene (CSM) are highly resistant to oxidizing media, such as chlorine bleaching liquor.

Rubber basis of rubber lining (Trade name[1])	Medium	Concentration	Temperature K (°C)	Resistance	References
a) Hard rubber lining					
a1) Factory rubber linings					
Natural rubber (NR) (Chemonit® 3B, EC Duro-Bond® Rubber, Vulkodurit® D3, Wagunit® H 1000)	aqueous CO_2 solution	saturated	≤ 355 (≤ 82)	+	[258]
		saturated	293–373 (20–100)	+	[259]
		n. d.	≤ 366 (≤ 93)	+	[260, 261]
	CO_2 gas, moist	n. d.	≤ 352 (≤ 79)	+	[260, 261]
	CO_2 gas	technically pure	293–313 (20–40)	+	[262]
		technically pure	353 (80)	−	[262]
	CO_2 gas, dry	technically pure	≤ 352–353 (≤ 79–80)	+	[260, 261, 263]
NR/Isoprene rubber (IR) (Wagunit® H 1118)	CO_2 gas	technically pure	293–373 (20–100)	+	[262]

+ = resistant   ⊕ = moderately resistant   − = not resistant
n. d. = not defined
[1] Trade names nonbinding

Table 60: Resistance of semi-hard and hard rubber linings to aqueous carbon dioxide solutions and carbon dioxide

Table 60: Continued

Rubber basis of rubber lining (Trade name[1])	Medium	Concentration	Temperature K (°C)	Resistance	References
NR/styrene-butadiene rubber (SBR) (Vulkodurit® 1250, Wagunit® 1050)	aqueous CO_2 solution	saturated	293–373 (20–100)	+	[259]
	CO_2 gas	technically pure	283–373 (10–100)	+	[262]
Isoprene rubber (IR) (Genakor® 022, Vulcoferran® 2190)	aqueous CO_2 solution	saturated	293–373 (20–100)	+	[259]
IR/SBR (Chemonit® 34)	CO_2 gas, dry	technically pure	≤ 353 (≤ 80)	+	[263]
SBR (Chemonit 10)	CO_2 gas, dry	technically pure	≤ 353 (≤ 80)	+	[263]
SBR/NR (Wagunit® H 1010)	CO_2 gas	technically pure	293–353 (20–80)	+	[262]
		technically pure	373 (100)	–	[262]
a2) Construction site rubber linings					
IR (Genakor® 022R)	aqueous CO_2 solution	saturated	293–373 (20–100)	+	[264]
IR/SBR (Chemonit® 34HW, Vulcoferran® 2194)	aqueous CO_2 solution	saturated	293–353 (20–80)	+	[264]
	CO_2 gas, dry	technically pure	353 (80)	⊕	[263]
NR (Chemonit® 31 HW, Kerabonit® D3 HW, Wagunit® H 1122)	aqueous CO_2 solution	saturated	293–353 (20–80)	+	[264]
	CO_2 gas, dry	technically pure	293–353 (20–80)	+	[262, 263]
NR/SBR, 60 Shore D (Wagunit® H 1110)	CO_2 gas	technically pure	293–353 (20–80)	+	[262]

+ = resistant ⊕ = moderately resistant – = not resistant
n. d. = not defined
[1] Trade names nonbinding

Table 60: Resistance of semi-hard and hard rubber linings to aqueous carbon dioxide solutions and carbon dioxide

Table 60: Continued

Rubber basis of rubber lining (Trade name[1])	Medium	Concentration	Temperature K (°C)	Resistance	References
b) Semi-hard rubber linings					
NR	aqueous CO_2 solution	saturated	≤ 355 (≤ 82)	+	[258]
		saturated	≤ 355 (≤ 82)	+ to ⊕	[231]
		n. d.	≤ 366 (≤ 93)	+	[261]
	CO_2 gas, moist	saturated	≤ 355 (≤ 82)	+ to ⊕	[231]
		n. d.	≤ 352 (≤ 79)	+	[261]
	CO_2 gas, dry	technically pure	≤ 352 (≤ 79)	+	[261]
		technically pure	≤ 355 (≤ 82)	+ to ⊕	[231]
SBR	aqueous CO_2 solution	saturated	≤ 355 (≤ 82)	+ to ⊕	[231]
	CO_2 gas, moist	saturated	≤ 355 (≤ 82)	+ to ⊕	[231]
	CO_2 gas, dry	technically pure	≤ 355 (≤ 82)	+ to ⊕	[231]

+ = resistant ⊕ = moderately resistant – = not resistant
n. d. = not defined
[1] Trade names nonbinding

Table 60: Resistance of semi-hard and hard rubber linings to aqueous carbon dioxide solutions and carbon dioxide

As shown by the resistance values in Table 60, factory-fabricated hard rubber linings are resistant to aqueous carbon dioxide solutions or carbon dioxide gas (dry or moist) up to 353 to 373 K (80 to 100°C). The hard rubber linings vulcanized with hot water on construction sites can be used up to 353 K (80°C). Frequently, graphite-filled rubber linings reach a slightly higher resistance than rubber linings containing mineral fillers or rubber powder.

Apart from the hard rubber linings produced with rubber sheets also a liquid hard rubber lining (trade name: Liquidline® 100) is offered on the market during the manufacture of which a liquid rubber layer is applied to the components in a spraying process and which is cured in the autoclave [265]. For Liquidline® 100 a maximum application temperature of 373 K (100°C) is indicated if exposed to saturated aqueous carbon dioxide solution [259]. Apart from the hard rubber linings, a maximum application temperature of 333 K (60°C) is indicated for a duroplastic lining on a phenol resin basis (trade name: Bornumharz® 6101) if exposed to saturated aqueous carbon dioxide solution, with mechanical properties being similar to those of hard rubber linings [259].

Besides the factory-fabricated soft rubber linings, soft rubber linings are widely applied on construction sites, both pre-vulcanized rubber sheets and cold-vulcanizing rubber sheets being used in the manufacturers' plants.

Rubber basis of soft rubber lining (Trade name[1])	Medium	Concentration	Temperature K (°C)	Resistance	References
a) Factory rubber linings					
Butyl rubber (IIR) (Vulkodurit® 1755)	aqueous CO_2 solution	saturated	≤ 339 (≤ 66)	+	[258]
		saturated	293–373 (20–100)	+	[259]
		n. d.	≤ 358 (≤ 85)	+	[261]
	CO_2 gas, moist	n. d.	≤ 352 (≤ 79)	+	[261]
	CO_2 gas, dry	technically pure	≤ 352 (≤ 79)	+	[261]
Bromobutyl rubber (BIIR) (Vulcoferran® 2208, Wagulast® BIIR 1642)	aqueous CO_2 solution	saturated	293–383 (20–110)	+	[259]
	CO_2 gas	technically pure	293–373 (20–100)	+	[262]
Chlorobutyl rubber (CIIR) (EC Duro-Bond® Chlorbutyl, Enduraflex VE621BC)	aqueous CO_2 solution	saturated	≤ 333 (≤ 60)	+ to ⊕	[231]
		saturated	≤ 355 (≤ 82)	+	[258, 260]
		saturated	≤ 361 (≤ 88)	+	[266]
	CO_2 gas, moist	saturated	≤ 333 (≤ 60)	+ to ⊕	[231]
		saturated	≤ 355 (≤ 82)	+	[260]
Chlorosulfonated polyethylene (CSM) (EC Duro-Bond Hypalon®, Vulkodurit® 1691-75, Wagulast® CSM 1720)	aqueous CO_2 solution	saturated	293–333 (20–60)	+	[259]
		n. d.	≤ 366 (≤ 93)	+	[260, 261]
	CO_2 gas, moist	saturated	≤ 333 (≤ 60)	+ to ⊕	[231]
		n. d.	≤ 366 (≤ 93)	+	[260, 261]
	CO_2 gas	technically pure	293–313 (20–40)	+	[262]
		technically pure	353 (80)	−	[262]
		technically pure	≤ 366 (≤ 93)	+	[260, 261]

+ = resistant   ⊕ = moderately resistant   − = not resistant
n. d. = not defined
[1] Trade names nonbinding

Table 61: Resistance of soft rubber linings to aqueous carbon dioxide solutions and carbon dioxide gas

Table 61: Continued

Rubber basis of soft rubber lining (Trade name[1])	Medium	Concentration	Temperature K (°C)	Resistance	References
Nitrile rubber (NBR) (Wagulast® NBR 1842)	aqueous CO_2 solution	saturated	≤ 339 (≤ 66)	+	[258]
	CO_2 gas	technically pure	293–313 (20–40)	+	[262]
		technically pure	353 (80)	–	[262]
Chloroprene rubber (CR) (EC Duro-Bond Neoprene®, Vulcoferran® 2503)	aqueous CO_2 solution	saturated	293–303 (20–30)	+	[259]
		saturated	≤ 355 (≤ 82)	+	[258]
		n. d.	≤ 366 (≤ 93)	+	[260, 261]
	CO_2 gas, moist	saturated	≤ 333 (≤ 60)	+ to ⊕	[231]
		n. d.	≤ 366 (≤ 93)	+	[260, 261]
	CO_2 gas, dry	technically pure	≤ 333 (≤ 60)	+ to ⊕	[231]
		technically pure	≤ 366 (≤ 93)	+	[260, 261]
NR/SBR/Butadiene caoutchouc (BR) (Wagulast® NR 1358)	CO_2 gas	technically pure	293–313 (20–40)	+	[262]
		technically pure	353 (80)	–	[262]
NR	aqueous CO_2 solution	saturated	≤ 333 (≤ 60)	+ to ⊕	[231]
		saturated	≤ 355 (≤ 82)	+	[258]
		n. d.	≤ 339 (≤ 66)	+	[261]
	CO_2 gas, moist	saturated	≤ 333 (≤ 60)	+ to ⊕	[231]
		n. d.	≤ 339 (≤ 66)	+	[261]
	CO_2 gas, dry	technically pure	≤ 333 (≤ 60)	+ to ⊕	[231]
			≤ 339 (≤ 66)	+	[261]
b) Construction site rubber linings with pre-vulcanized rubber sheets					
Chlorobutyl rubber (CIIR) (Kerabutyl® BS)	aqueous CO_2 solution	saturated	293–333 (20–60)	+	[264]
CIIR, double-layer sheet (Kerabutyl® V)	aqueous CO_2 solution	saturated	293–353 (20–80)	+	[264]

+ = resistant ⊕ = moderately resistant – = not resistant
n. d. = not defined
[1] Trade names nonbinding

Table 61: Resistance of soft rubber linings to aqueous carbon dioxide solutions and carbon dioxide gas

Table 61: Continued

Rubber basis of soft rubber lining (Trade name[1])	Medium	Concentration	Temperature K (°C)	Resistance	References
c) Construction site rubber linings with cold-vulcanizing rubber sheets					
BIIR (Kerabutyl® BB-S, Vulcoferran® 2208, Wagulast® BIIR 1643)	aqueous CO_2 solution	saturated	293–353 (20–80)	+	[264]
		saturated	293–373 (20–100)	+	[264]
	CO_2 gas	technically pure	293–353 (20–80)	+	[262]
NBR (Wagulast® NBR 1833)	CO_2 gas	technically pure	293–313 (20–40)	+	[262]
		technically pure	353 (80)	–	[262]
CR (Kerapren® VB, Wagulast® CR 1504)	aqueous CO_2 solution	saturated	293–313 (20–40)	+	[264]
	CO_2 gas	technically pure	293–313 (20–40)	+	[262]
		technically pure	353 (80)	–	[262]
CSM (Wagulast® CSM 1717)	CO_2 gas	technically pure	293–313 (20–40)	+	[262]
		technically pure	353 (80)	–	[262]

+ = resistant   ⊕ = moderately resistant   – = not resistant
n. d. = not defined
[1] Trade names nonbinding

Table 61: Resistance of soft rubber linings to aqueous carbon dioxide solutions and carbon dioxide gas

If exposed to aqueous carbon dioxide solutions soft rubber linings on the basis of the butyl rubbers IIR, CIIR or BIIR can be used at temperature up to 353 to 373 K (80 to 100°C) and in single cases even up to 383 K (110°C). Regarding soft rubber linings on the basis of CR, CSM and NR it should be noted that the water vapor permeability strongly increases at temperatures above about 333 K (60°C); this fact, therefore, needs to be considered with caution in connection with the high application temperatures for these rubber linings indicated in [258, 260, 261].

Thermoplastic linings

Thermoplastic layer (medium side) (Trade name[1])	Back layer (adhesive side)	Thickness Thermoplastic layer mm	Medium	Concentration	Temperature K (°C)	Resistance	References
a) Thermoplastic webs glued to the entire surface							
Polypropylene (PP) (EC Duro-Bond® PP Lining)	soft rubber	2.3; 3.0	CO_2 gas	100%	295–373 (22–100)	+	[260]
Polyvinyl chloride with plasticizers (PVC-P) (Trovidur® W 2000)		2.3; 3.0; 4.6	aqueous CO_2 solution	saturated	≤ 323 (≤ 50)	+	[267]
Ethylene chlorotrifluoroethylene copolymer (ECTFE) (EC Duro-Bond® E-CTFE Lining)	fabric or soft rubber	1.5; 2.3	aqueous CO_2 solution	n. d.	294–380 (21–107)	+	[260]
		1.5; 2.3	CO_2 gas, moist	n. d.	294–380 (21–107)	+	[260]
		1.5; 2.3	CO_2 gas, dry	technically pure	294–380 (21–107)	+	[260]
Ethylene tetrafluoroethylene copolymer (ETFE) (EC Duro-Bond® ETFE Lining)	fabric	1.5; 2.3	aqueous CO_2 solution	n. d.	≤ 383 (≤ 110)	+	[260]
		1.5; 2.3	CO_2 gas, moist	n. d.	≤ 383 (≤ 110)	+	[260]
		1.5; 2.3	CO_2 gas, dry	technically pure	≤ 383 (≤ 110)	+	[260]
Polyvinylidene fluoride (PVDF) (EC Duro-Bond® PVDF Lining)	fabric or soft rubber	1.5; 2.3; 3.0	CO_2 gas, moist	n. d.	294–380 (21–107)	+	[260]
		1.5; 2.3; 3.0	CO_2 gas, dry	technically pure	294–380 (21–107)	+	[260]
Polytetrafluoroethylene, modified (PTFE-M) (EC Duro-Bond® PTFE-M Lining)	fiber mesh	2.0; 3.0	aqueous CO_2 solution	any	≤ 383 (≤ 110)	+	[260]
		2.0; 3.0	CO_2 gas, moist	any	≤ 383 (≤ 110)	+	[260]
		2.0; 3.0	CO_2 gas, dry	any	≤ 383 (≤ 110)	+	[260]

+ = resistant
n. d. = not defined; 1) Trade names nonbinding

Table 62: Resistance of thermoplastic linings to aqueous carbon dioxide solutions and carbon dioxide gas

Table 62: Continued

Thermoplastic layer (medium side) (Trade name[1])	Back layer (adhesive side)	Thickness Thermoplastic layer mm	Medium	Concentration	Temperature K (°C)	Resistance	References
Tetrafluoroethylene-perfluoropropyl-vinylether copolymer (PFA) (EC Duro-Bond® PFA Lining)	fiber mesh	1.5; 2.3	aqueous CO_2 solution	any	≤ 383 (≤ 110)	+	[260]
		1.5; 2.3	CO_2 gas, moist	any	≤ 383 (≤ 110)	+	[260]
		1.5; 2.3	CO_2 gas, dry	any	≤ 383 (≤ 110)	+	[260]
Tetrafluoroethylene-hexafluoropropylene copolymer (FEP) (EC Duro-Bond® FEP Lining)	fiber mesh	1.5; 2.3	aqueous CO_2 solution	any	≤ 383 (≤ 110)	+	[260]
		1.5; 2.3	CO_2 gas, moist	any	≤ 383 (≤ 110)	+	[260]
		1.5; 2.3	CO_2 gas, dry	any	≤ 383 (≤ 110)	+	[260]
Tetrafluoroethylene/perfluoromethyl vinyl ether copolymer (MFA) (EC Duro-Bond® MFA Lining)	fiber mesh	1.5; 2.3	aqueous CO_2 solution	any	≤ 383 (≤ 110)	+	[260]
		1.5; 2.3	CO_2 gas, moist	any	≤ 383 (≤ 110)	+	[260]
		1.5; 2.3	CO_2 gas, dry	any	≤ 383 (≤ 110)	+	[260]
b) Thermoplastic webs mechanically anchored in concrete							
Polyethylene (PE) (Anchor-Lok® PE)		3; 5	aqueous CO_2 solution	n. d.	293–333 (20–60)	+	[233]
		3; 5	CO_2 gas	technically pure	293–333 (20–60)	+	[233]
Polypropylene (PP) (Anchor-Lok® PP)		3; 5	aqueous CO_2 solution	n. d.	293–333 (20–60)	+	[233]
		3; 5	CO_2 gas	technically pure	293–333 (20–60)	+	[233]

+ = resistant
n. d. = not defined; 1) Trade names nonbinding

Table 62: Resistance of thermoplastic linings to aqueous carbon dioxide solutions and carbon dioxide gas

Table 62: Continued

Thermoplastic layer (medium side) (Trade name[1])	Back layer (adhesive side)	Thickness Thermo-plastic layer mm	Medium	Concentration	Temperature K (°C)	Resistance	References
Polyvinyl chloride (PVC-U) (Anchor-Lok® PVC)		3; 5	aqueous CO_2 solution	n. d.	293–333 (20–60)	+	[233]
		3; 5	CO_2 gas	technically pure	293–333 (20–60)	+	[233]
Polyvinylidene fluoride (PVDF) (Anchor-Lok® PVDF)		3; 5	aqueous CO_2 solution	n. d.	293–373 (20–100)	+	[233]
		3; 5	CO_2 gas	technically pure	293–373 (20–100)	+	[233]

+ = resistant
n. d. = not defined; 1) Trade names nonbinding

Table 62: Resistance of thermoplastic linings to aqueous carbon dioxide solutions and carbon dioxide gas

The resistance values for the thermoplastic sheets mechanically anchored in concrete mainly correspond to the values indicated in Section C for the relevant thermoplastics. The upper application temperature of glued-down thermoplastic sheets on the basis of perfluoropolymers is determined by the limited thermal resistance of the used epoxy resin glue. The application range of these thermoplastic sheets could be expanded to include higher temperatures by using a glue with a higher thermal resistance.

D 2 Seals and packings

The following information is include flat seals, packings, O-rings and sliding ring seals.

Flat gaskets

According to the standard DIN 28091 Parts 1–4 [268] the flat seals are classified into several groups as shown in Table 63. The pressure/temperature diagram showing the permissible temperature and pressure ranges for utilization of the material is useful when selecting suitable flat seals for a particular application. Such pT diagrams are essentially based on practical experience. As example Figure 84 shows a pT diagram for a flat seal made of graphite, Aramid fibers and NBR elastomer (novatec® PREMIUM II) with carbon dioxide gas (100 %) as medium [269]. It is seen that the utilization temperature range becomes narrower with increasing operating pressure. With an operating pressure of e.g. 20 bar the seal can be utilized up to a temperature of about 473 K (200°C), but with a pressure of 60 bar only from up to about 423 K (150°C).

Instead of the pT diagrams, maximum values for the product of operating pressure and temperature can also be specified, that must not be exceeded, whereby additional separate specifications of the maximum operating temperature and the maximum operating pressure are made for the individual seals [270].

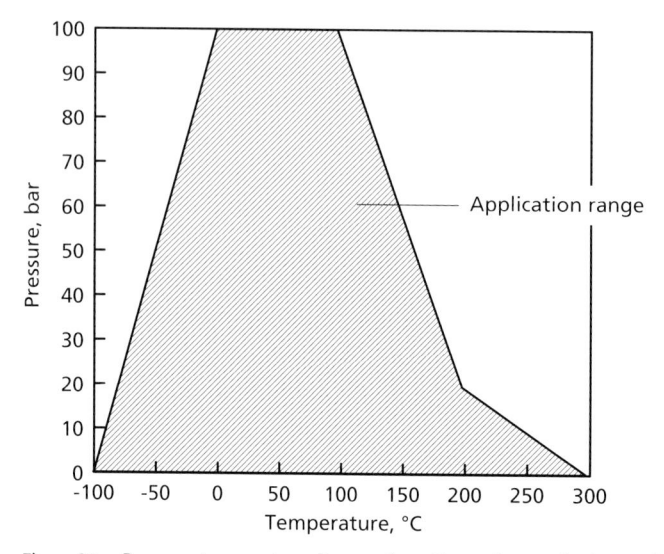

Figure 84: Pressure-temperature diagram for a flat gasket on the basis of graphite/aramid fibers/NBR (novatec® PREMIUM II) and carbon dioxide gas a medium [269]

For the final selection of a seal, apart from the chemical durability, further key parameters described in the standard DIN 28090-1 [271] must be taken into consideration. These are, for example, the minimum and maximum specific surface pressure and the sealing class. The manufacturers provide calculating programs for designing the flat seals, in which the relevant parameters such as the dimensions, the sealing class and the operating parameters are taken into consideration.

As can be seen from Table 63, flat gaskets with fibers can be used up to 443 K (170°C) and flat gaskets on the basis of PTFE can be used up to 523 K (250°C) if exposed to aqueous carbon dioxide solutions or moist carbon dioxide. Flat seals on the basis of expanded graphite feature extremely high utilization temperatures and very broad resistance to chemicals.

Flat gasket components (Trade name[1])	Medium	Concentration	Maximum application temperature K (°C)	Operating pressure	References
a) Flat gaskets on the basis of fibers/fillers/binding agents					
Carbon fibers/fillers/ nitrile rubber (NBR) (Klingersil® C-4500)	CO_2 gas	technically pure	523 (250)	max. 40 bar	[272]
Synthetic fibers/NBR (Klingersil® C-4430)	CO_2 gas	technically pure	523 (250)	max. 40 bar	[272]
Aramid fiber/graphite/NBR (novatec® PREMIUM II)	aqueous CO_2 solution	n. d.	443 (170)	20 bar	[269]
	CO_2 gas	technically pure	473 (200)	20 bar	[269]
Aramid fiber/filler/elastomer (AFM 34)	CO_2 gas	technically pure	523 (250)		[273]
b) Flat gaskets on the basis of polytetrafluoroethylene (PTFE)					
PTFE/special fillers (Klingertop-chem®-2000, novaflon® 200)	aqueous CO_2 solution	n. d.	503 (230)	40 bar	[269]
	CO_2 gas	technically pure	513 (240)	20 bar	[269]
		technically pure	523 (250)	max. 60 bar	[272]
PTFE Gylon® Standard, Gylon® Blau, Gylon® Weiß	CO_2 gas, moist	n. d.	523 (250)	max. 45 bar	[270]
	CO_2 gas, dry	technically pure	523 (250)	max. 45 bar	[270]
PTFE, multidirectionally expanded (novaflon® 500)	aqueous CO_2 solution	n. d.	513 (240)	40 bar	[269]
	CO_2 gas	technically pure	523 (250)	20 bar	[269]
c) Flat gaskets on the basis of graphite					
Graphite, expanded/synthetic fibers (Klingertop-graph®-2000)	CO_2 gas	technically pure	573 (300)	max. 40 bar	[272]
Graphite, expanded and precompressed (novaphit® VS)	aqueous CO_2 solution	n. d.	523 (250)	50 bar	[269]
	CO_2 gas	technically pure	613 (340)	50 bar	[269]
Graphite, expanded/expansion metal made of stainless steel 1.4404 (novaphit® SSTC)	aqueous CO_2 solution	n. d.	623 (350)	180 bar	[269]
	CO_2 gas	technically pure	703 (430)	180 bar	[269]

[1] Trade names nonbinding

Table 63: Maximum application temperatures of flat gaskets for aqueous carbon dioxide solutions and carbon dioxide gas

As a strip sealing for metallic device flanges a sealing strip of polytetrafluoroethylene (PTFE) (trade name: Gore-Tex® Series 300 gasket tape) is offered, which is applicable to aqueous carbon dioxide solutions and carbon dioxide gas (moist or dry) up to a maximum of 523 K (250°C) and 40 bar. For enamelled device flange connections another PTFE sealing strip (trade name: Gore-Tex® Series 600 gasket tape) is recommended for applications of up to 493 K (220° C) and 6 bar [274].

Packings

Packings or compression packings used for sealing rotating machine components, e.g. in pumps, mixers, stirrers or fittings, consist of braided yarns provided with an impregnation or filling. The yarns may consist of ramie-, polyacrylnitrile (PAN) or aramid fibers or may be produced on the basis of carbon, graphite or PTFE. PTFE or graphite is frequently utilized for impregnation and filling. When selecting the suitable packing for a particular application, the speed of the moving machine parts, the operating pressure and temperature range and the chemical stress must be taken into primary consideration. Ramie fiber packings are not recommended for an exposure to aqueous carbon dioxide solutions above 373 K (100°C); otherwise the wide range of products of the above yarns and impregnations may be used [275–277]. With regard to the continuous operating temperatures it is advisable to consult the manufacturer, because for each particular application the pressure and machine speed as well as other parameters have to be taken into consideration.

O-rings

Chiefly elastomers as described in Section C are utilized for manufacturing O-rings. When selecting suitable elastomers for a particular application, other technical properties such as e.g. abrasion resistance, permeability or deformation under pressure can be of importance in addition to the mechanical properties, the utilization temperature range and the stress source media.

Details regarding the resistance of the elastomers to aqueous carbon dioxide solutions and carbon dioxide gas have been dealt with in Section C under "Elastomers". As supplement thereto, Table 64 contains durability data from manufacturers of O-rings (and sealing rings with other designations). As can be seen, sealing materials on the basis of ACM, AU and EU are not recommended. All other elastomer materials are suited for ambient temperatures; for FFKM elastomers an upper application temperature of 503 K (230°C) is indicated.

Elastomer basis	Medium	Concentration	Temperature K (°C)	Resistance	References
Natural rubber (NR)	aqueous CO_2 solution	n. d.	RT	+	[278]
	CO_2 gas, moist	n. d.	RT	+	[278, 279]
	CO_2 gas, dry	technically pure	RT	+	[278]
		technically pure	333 (60)	+	[279]
Styrene butadiene rubber (SBR)	aqueous CO_2 solution	n. d.	RT	+	[278]
	CO_2 gas, moist	n. d.	RT	+	[278, 279]
	CO_2 gas, dry	technically pure	RT	+	[278]
		technically pure	333 (60)	+	[279]
Chloroprene rubber (CR)	aqueous CO_2 solution	n. d.	RT	+	[278]
		n. d.	RT	+ to ⊕	[280]
	CO_2 gas, moist	n. d.	RT	+	[278, 279]
		n. d.	RT	+ to ⊕	[280]
	CO_2 gas, dry	technically pure	RT	+	[278]
		technically pure	RT	+ to ⊕	[280]
		technically pure	333 (60)	+	[279]
Acrylonitrile butadiene rubber (Nitrile rubber) (NBR)	aqueous CO_2 solution	n. d.	RT	+	[278, 280]
	CO_2 gas, moist	n. d.	RT	+	[278–280]
	CO_2 gas, dry	technically pure	RT	+	[278, 280]
		technically pure	333 (60)	+	[279]
Hydrogenated nitrile rubber (HNBR)	aqueous CO_2 solution	n. d.	RT	+	[278, 280]
	CO_2 gas, moist	n. d.	RT	+	[278–280]
	CO_2 gas, dry	technically pure	RT	+	[278, 280]
		technically pure	333 (60)	+	[279]

+ = resistant   ⊕ = moderately resistant   − = not resistant
RT = room temperature, n. d. = not defined

Table 64: Resistance of elastomer materials for O-rings and other sealing rings to aqueous carbon dioxide solutions and carbon dioxide gas

Table 64: Continued

Elastomer basis	Medium	Concentration	Temperature K (°C)	Resistance	References
Isobutene isoprene rubber (Butyl rubber) (IIR)	aqueous CO_2 solution	n. d.	RT	+	[278]
	CO_2 gas, moist	n. d.	RT	+	[278, 279]
	CO_2 gas, dry	technically pure	RT	+	[278]
		technically pure	333 (60)	+	[279]
Ethylene propylene diene rubber (EPDM)	aqueous CO_2 solution	n. d.	RT	+	[278, 280]
	CO_2 gas, moist	n. d.	RT	+	[278, 279]
		n. d.	RT	+ to ⊕	[280]
	CO_2 gas, dry	technically pure	RT	+	[278]
		technically pure	RT	+ to ⊕	[280]
		technically pure	333 (60)	+	[279]
Chlorosulfonated polyethylene (CSM)	aqueous CO_2 solution	n. d.	RT	+	[278]
	CO_2 gas, moist	n. d.	RT	+	[278, 279]
	CO_2 gas, dry	technically pure	RT	+	[278]
		technically pure	333 (60)	+	[279]
Acrylate rubber (ACM)	aqueous CO_2 solution	n. d.	RT	−	[278, 280]
	CO_2 gas, moist	n. d.	293 (20)	+	[279]
		n. d.	RT	−	[278, 280]
	CO_2 gas, dry	technically pure	RT	+	[278]
		technically pure	RT	+ to ⊕	[280]
		technically pure	333 (60)	+	[279]
Fluorinated rubber (FKM)	aqueous CO_2 solution	n. d.	RT	+	[278, 280]
	CO_2 gas, moist	n. d.	RT	+	[278–280]
	CO_2 gas, dry	technically pure	RT	+	[278, 280]
		technically pure	333 (60)	+	[279]

+ = resistant   ⊕ = moderately resistant   − = not resistant
RT = room temperature, n. d. = not defined

Table 64: Resistance of elastomer materials for O-rings and other sealing rings to aqueous carbon dioxide solutions and carbon dioxide gas

Table 64: Continued

Elastomer basis	Medium	Concentration	Temperature K (°C)	Resistance	References
Perfluoro rubber (FFKM)	aqueous CO$_2$ solution	any	≤ 503 (≤ 230)	+	[281]
		n. d.	RT	+	[278]
	CO$_2$ gas, moist	n. d.	RT	+	[278, 279]
	CO$_2$ gas, dry	technically pure	RT	+	[278]
		technically pure	333 (60)	+	[279]
		technically pure	≤ 503 (≤ 230)	+	[281]
Silicone rubber (MQ, VMQ)	aqueous CO$_2$ solution	n. d.	RT	+	[278]
		n. d.	RT	+ to ⊕	[280]
	CO$_2$ gas, moist	n. d.	RT	+	[278, 279]
		n. d.	RT	+ to ⊕	[280]
	CO$_2$ gas, dry	technically pure	RT	+	[278]
		technically pure	RT	+ to ⊕	[280]
		technically pure	333 (60)	+	[279]
Fluorosilicone rubber (FVMQ)	aqueous CO$_2$ solution	n. d.	RT	+	[278]
		n. d.	RT	+ to ⊕	[280]
	CO$_2$ gas, moist	n. d.	RT	+	[278, 279]
		n. d.	RT	+ to ⊕	[280]
	CO$_2$ gas, dry	technically pure	RT	+	[278]
		technically pure	RT	+ to ⊕	[280]
		technically pure	333 (60)	+	[279]
Polyurethane rubber (AU, EU)	aqueous CO$_2$ solution	n. d.	RT	+ to ⊕	[280]
		n. d.	RT	–	[278]
	CO$_2$ gas, moist	n. d.	RT	–	[278, 280]
	CO$_2$ gas, dry	technically pure	RT	+	[278]
		technically pure	RT	–	[280]

+ = resistant ⊕ = moderately resistant – = not resistant
RT = room temperature, n. d. = not defined

Table 64: Resistance of elastomer materials for O-rings and other sealing rings to aqueous carbon dioxide solutions and carbon dioxide gas

[278] particularly recommends for O-rings an elastomer on EPDM basis if exposed to aqueous carbon dioxide solutions and an elastomer on NBR basis if exposed to (moist or dry) carbon dioxide gas.

As elaborated in Section C under "Elastomers" FKM elastomers are not recommended for applications with an exposure to carbon dioxide under high pressure.

For applications where explosive decompressions may occur as a result of a sudden pressure drop, [282] recommends an FFKM elastomer (trade name: Isolast® J9510).

Mechanical seals (Sliding ring seals)

Examples for the application of mechanical seals are contained in Table 65 [283].

Details of mechanical seals	Design / Material
a) Aqueous carbon dioxide solution (not saturated) up to about 363 K (90°C)	
Shaft seal assembly	single seal, internal assembly
Auxiliary pipeline	circulation of the liquid from the pump exit to the shaft seal room (with internal backflow)
Construction type of the sliding ring seal	sliding ring seal with O-ring subsidiary seals
Material of sliding ring	special chrome-molybdenum casting
Material of counter ring	carbon graphite, impregnated with synthetic resin
Material of subsidiary seal	elastomer on the basis of nitrile rubber (NBR)
Material of spring	CrNiMo steel1.4571
b) Carbon dioxide up to 333 K (60°C)	
Shaft seal assembly	double seal, back to back configuration
Auxiliary pipeline	sealing or quenching liquid of a pressure vessel; circulation by the thermosiphon effect or pump
Construction type of the mechanical seals	mechanical seals with O-ring subsidiary seals
Material of sliding ring	special chrome-molybdenum casting
Material of counter ring	carbon graphite, impregnated with synthetic resin
Material of subsidiary seal	elastomer on the basis of nitrile rubber (NBR)
Material of spring	CrNiMo steel 1.4571

[1] The sealing pressure must be at least 3 bar higher than the vapor pressure of CO_2.

Table 65: Examples of mechanical seal designs for exposure to aqueous carbon dioxide solutions or gaseous or liquefied carbon dioxide [283]

Table 65: Continued

Details of mechanical seals	Design / Material
c) Carbon dioxide, liquefied[1]	
Shaft seal assembly	single seal internal assembly; seal with subsidiary seal or packing
Auxiliary pipeline	circulation of the liquid from the pump exit to the shaft seal room (with internal backflow)
Construction type of the mechanical seals	special construction
Material of sliding ring	silicon carbide (SiC), pressureless sintered
Material of counter ring	carbon graphite, antimony impregnated
Material of subsidiary seal	elastomer on the basis of nitrile rubber (NBR)
Material of spring	CrNiMo steel 1.4571

[1] The sealing pressure must be at least 3 bar higher than the vapor pressure of CO_2.

Table 65: Examples of mechanical seal designs for exposure to aqueous carbon dioxide solutions or gaseous or liquefied carbon dioxide [283]

D 3 Composite materials

Glass fiber reinforced plastics (GFRP)

Duroplastic glass fiber reinforced plastics (GFRP) are produced from reaction resins and glassfibers as a reinforcing material, for pipes and equipment mainly unsaturated polyester (UP) and vinyl ester (VE) resins as well as epoxy (EP) resins being used. The glass fibers are applied as rovings, settings, fabrics, mats or complexes as well as fleeces (for surface layers). The glass fibers usually consist of E-glass, a low alkali aluminium-boron-silicate glass. For some time also reinforcing materials of boron-free aluminium silicate glass types with improved chemical resistance are available, in particular Advantex® glass (former name: ECR glass) and E-CR glass, both exhibiting a similar chemical resistance. C-glass is available on the market only as fleeces. The reinforcement materials made of glass fibers (alternative designation: reinforcing materials manufactured from textile glass), the structure, the parameters of UP and VE resins and the manufacture and parameters of laminates of glass fiber reinforced UP and VE resins are standardized.

Details regarding the resistance of laminates and components of glass fiber reinforced UP, VE and EP resins to aqueous carbon dioxide solutions and carbon dioxide gas are contained in Table 66. Moreover, it also contains resistance values for carbon fiber-reinforced VE resins and glass fiber reinforced polyphenylene sulfide (PPS).

Reaction resin (Trade name[1])	Medium	Concentration	Temperature K (°C)	Resistance	References
a) Glass fiber reinforced plastics (GFRP) – laminates					
UP resin on the basis of alcoxylated bisphenol A (Atlac® 382)	aqueous CO_2 solution	any	≤ 353 (≤ 80)	+	[284]
	CO_2 gas, dry	technically pure	≤ 453 (≤ 180)	+	[284]
UP resin on the basis of orthophthalic acid/standard glycolene (Palatal® P 69)	aqueous CO_2 solution	any	≤ 313 (≤ 40)	+	[284]
	CO_2 gas, dry	technically pure	≤ 333 (≤ 60)	+	[284]
UP resins based on isophthalic acid/neopentyl glycol (Aropol® 7241, Palatal® A 410)	aqueous CO_2 solution	any saturated	≤ 333 (≤ 60) ≤ 344 (≤ 71)	+ +	[284] [285]
	CO_2 gas, moist	100 %	≤ 366 (≤ 93)	+	[285]
	CO_2 gas, dry	technically pure	≤ 373 (≤ 100)	+	[284]
UP resin based on HET acid (Hetron® 197-3)	aqueous CO_2 solution	saturated	≤ 344 (≤ 71)	+	[285]
	CO_2 gas, moist	100 %	≤ 394 (≤ 121)	+	[285]
VE resins based on bisphenol A (Atlac® 430, Derakane® 411, Hetron® 922)	aqueous CO_2 solution	any saturated	≤ 353 (≤ 80) ≤ 344 (≤ 71)	+ +	[284] [285]
	CO_2 gas, moist	100 %	≤ 372 (≤ 99)	+	[285]
	CO_2 gas, dry	any technically pure	≤ 438 (≤ 165) ≤ 373 (≤ 100)	+ +	[286] [284]
VE resins based on Novolak (Atlac® 590, Derakane® 470, Nil-Cor® 310)	aqueous CO_2 solution	any	≤ 353 (≤ 80)	+	[284]
	CO_2 gas, dry	any technically pure technically pure	≤ 478 (≤ 205) ≤ 394 (≤ 121) ≤ 473 (≤ 200)	+ + +	[286] [287] [284]

+ = resistant   ⊕ = moderately resistant
UP = unsaturated polyester
HET acid = hexachloro-endomethylene-tetrahydrophthalic acid
VE = vinyl ester
EP = epoxide
n. d. = not defined
RT = room temperature
[1] Trade names nonbinding

Table 66: Resistance of fiber-reinforced plastics to aqueous carbon dioxide solutions and carbon dioxide gas

Table 66: Continued

Reaction resin (Trade name[1])	Medium	Concentration	Temperature K (°C)	Resistance	References
VE resin, urethane modified (Atlac® 580)	aqueous CO_2 solution	any	≤ 353 (≤ 80)	+	[284]
	CO_2 gas, dry	technically pure	≤ 423 (≤ 150)	+	[284]
EP resin/aromatic polyamine (AVT 530, Nil-Cor® 610XP)	aqueous CO_2 solution	any	≤ 355 (≤ 82)	+	[288]
	CO_2 gas, dry	technically pure	≤ 394 (≤ 121)	+	[288]
		technically pure	≤ 408 (≤ 135)	+	[287]
Polyphenylene sulfide (PPS) (Nil-Cor® 510)	CO_2 gas, dry	technically pure	≤ 366 (≤ 93)	+	[287]
b) Glass fiber reinforced plastics (GFRP) – components					
UP resin on the basis of orthophthalic acid Pipes	CO_2 gas	technically pure	≤ 323 (≤ 50)	+	[289]
UP resin on the basis of isophthalic acid Pipes, pultruded GFRP parts	aqueous CO_2 solution	n. d.	RT-339 (66)	+	[290]
	CO_2 gas	technically pure	RT-339 (66)	+	[290]
		technically pure	≤ 333–353 (≤ 60–80)	+	[289]
VE resin Gratings, pultruded GFRP parts	aqueous CO_2 solution	n. d.	RT-344 (71)	+	[290]
	CO_2 gas	technically pure	RT-344 (71)	+	[290]
		technically pure	≤ 372 (≤ 99)	+	[291]
VE resin, pipes with a resin rich inner layer 0.5 mm (Fiberdur® VE)	aqueous CO_2 solution	n. d.	298–368 (25–95)	+	[292]
	CO_2 gas	technically pure	298–368 (25–95)	+	[292]
		technically pure	≤ 368–393 (≤ 95–120)	+	[289]

+ = resistant ⊕ = moderately resistant
UP = unsaturated polyester
HET acid = hexachloro-endomethylene-tetrahydrophthalic acid
VE = vinyl ester
EP = epoxide
n. d. = not defined
RT = room temperature
[1] Trade names nonbinding

Table 66: Resistance of fiber-reinforced plastics to aqueous carbon dioxide solutions and carbon dioxide gas

Table 66: Continued

Reaction resin (Trade name[1])	Medium	Concentration	Temperature K (°C)	Resistance	References
VE resin, pipes/components with chemical barrier 2.5 mm (Fiberdur® CSVE)	aqueous CO_2 solution	n. d.	298–368 (25–95)	+	[292]
	CO_2 gas	technically pure	298–368 (25–95)	+	[292]
EP resin/polyamine, hot-curing Pipes with a resin rich inner layer 0.5 mm (Fiberdur® EP, Wavistrong®)	aqueous CO_2 solution	n. d.	≤ 333 (≤ 60)	+	[293]
		n. d.	298–338 (25–65)	+	[292]
	CO_2 gas	technically pure	≤ 373 (≤ 100)	+	[293]
		technically pure	298–383 (25–110)	+	[292]
EP resin/polyamine, hot-curing Pipes with a chemical barrier 2.5 mm (Fiberdur® CSEP)	aqueous CO_2 solution	n. d.	298–338 (25–65)	+	[292]
	CO_2 gas	technically pure	298–393 (25–120)	+	[292]
		technically pure	≤ 397 (≤ 124)	+	[294]
EP resin/polyamine, hot-curing Centrifugally cast pipes (Centricast® II)	aqueous CO_2 solution	n. d.	≤ 338 (≤ 65)	+	[292]
	CO_2 gas	technically pure	≤ 383 (≤ 110)	+	[292]
EP resin/anhydride hardener Pipes	aqueous CO_2 solution	n. d.	322–372 (49–99)	⊕	[295]
	CO_2 gas	technically pure	322–372 (49–99)	⊕	[295]
c) Graphite fiber reinforced plastics					
VE resin (Nil-Cor® 300)	CO_2 gas	technically pure	≤ 450 (≤ 177)	+	[287]
Polyphenylene sulfide (PPS) (Nil-Cor® 500)	CO_2 gas, dry	technically pure	≤ 408 (≤ 135)	+	[287]

+ = resistant   ⊕ = moderately resistant
UP = unsaturated polyester
HET acid = hexachloro-endomethylene-tetrahydrophthalic acid
VE = vinyl ester
EP = epoxide
n. d. = not defined
RT = room temperature
[1] Trade names nonbinding

Table 66: Resistance of fiber-reinforced plastics to aqueous carbon dioxide solutions and carbon dioxide gas

If exposed to aqueous carbon dioxide solutions or moist carbon dioxide, an upper application temperature of about 373 K (100°C) is indicated for components (e.g. pipes) of glass fiber reinforced VE resins and of 338 (65°C) for components of glass fiber reinforced EP resins hardened with aromatic polyamines. In contrast, GFRP pipes of EP resins and dicarbonic acid anhydride hardeners are only moderately resistant. If exposed to dry carbon dioxide, the upper application temperatures are clearly higher, e.g. 393 K (120°C) for EP-GFRP pipes.

In [296] the resistance of GFRP pipe specimens (length: 0.9 m; diameter: 51 mm; wall thickness: 1.8 mm) containing an epoxy, vinyl ester or an unsaturated polyester resin as the resin matrix, to (moist or dry) carbon dioxide gas, liquid carbon dioxide and H_2S containing carbon dioxide gas was examined at 297 K (24°C) and pressures from 20.7 to 58.6 bar over a period of up to 48 months by determining the annular tensile strength according to ASTM D2290 in various intervals. As can be seen from Table 67, the pipe specimens were able to withstand the chemical load under the test conditions applied.

Medium/pressure	Annular tensile strength retention in % after				
	12 months	18 months	24 months	36 months	48 months
a) GFRP pipes with an epoxy resin/aromatic polyamine matrix					
CO_2 gas, dry / 20.7 bar	88–109	100–104	94	97–98	96
CO_2 gas, dry / 41.4 bar	101	102			
CO_2 gas, moist / 20.7 bar	95		91	92	
CO_2 gas, moist / 41.4 bar	90		96	93	
Liquid CO_2 / 58.6 bar	110	104			
Gas mixture (95 % CO_2 and 5 % H_2S), moist / 41.4 bar	87–88				
b) GFRP pipes with a vinyl ester resin matrix					
CO_2 gas, moist / 20.7 bar	101				
CO_2 gas, moist / 41.4 bar	105				
Gas mixture (95 % CO_2 and 5 % H_2S), moist / 41.4 bar	86				
c) GFRP pipes with an unsaturated polyester resin matrix (isophthalic acid basis)					
CO_2 gas, moist / 20.7 bar	90				
CO_2 gas, moist / 41.4 bar	101				
Gas mixture (95 % CO_2 and 5 % H_2S), moist / 41.4 bar	102				

Initial annular tensile strength: 272 MPa (epoxy resin GFRP), 277 MPa (vinyl ester resin GFRP), 261 MPa (unsat. polyester resin GFRP)

Table 67: Annular tensile strength retention of GFRP pipes under carbon dioxide pressure load at 297 K (24°C) [296]

Thermoplastic GFRP composite materials

In the composite materials of thermoplastic sheets on the inner or medium side and GFRP laminates as the outer layer as are widely used in the production of pipes and chemical equipment for process plants, the GFRP laminates provide the necessary strength, bending stiffness and heat distortion resistance, whereas the thermoplastic layers are characterized by high density and chemical resistance, a very good abrasion resistance and a low solids adhesion tendency.

As internal lining materials chiefly PVC-U and the PP-types, but also PE-HD and PVC-C are utilized. For very high chemical and thermal loads, the partially or fully fluorinated thermoplastics PVDF, ECTFE, ETFE, FEP, MFA, PFA or PTFE (modified) are considered.

In the case of composite materials, the selection of the UP or VP resin for the GFRP layers essentially depends on the operating temperature of the component. UP resins on the basis of orthophthalic acid are utilizable up to about 353 K (80°C), whereas UP resins based on isophthalic acid and terephthalic acids as well as VE resins can be utilized at higher temperatures (up to about 423 K (150°C) for the VE resin novolak type).

Thermoplastic material	Medium	Concentration	Temperature K (°C)	Resistance	References
Polyethylene (PE)	CO_2 gas, moist	technically pure	293–333 (20–60)	+	[297]
	CO_2 gas, dry	technically pure	293–333 (20–60)	+	[297]
Polypropylene (PP)	CO_2 gas, moist	technically pure	293–333 (20–60)	+	[297]
	CO_2 gas, dry	technically pure	293–353 (20–80)	+	[297]
Polyvinyl chloride, without plasticizers (PVC-U)	aqueous CO_2 solution	≤ saturated	≤ 313 (≤ 40)	+	[168]
	CO_2 gas, moist	technically pure	293–313 (20–40)	+	[297]
		technically pure	333 (60)	⊕	[297]
	CO_2 gas, dry	technically pure	293–333 (20–60)	+	[297]
Polyvinyl chloride, chlorinated (PVC-C)	CO_2 gas, moist	technically pure	293–353 (20–80)	+	[297]
	CO_2 gas, dry	technically pure	293–353 (20–80)	+	[297]
Tetrafluoroethylene-perfluoro-propylvinylether copolymer (PFA)	CO_2 gas, moist	technically pure	293–423 (20–150)	+	[297]
	CO_2 gas, dry	technically pure	293–423 (20–150)	+	[297]

+ = resistant   ⊕ = moderately resistant

Table 68: Resistance of thermoplastic GFRP composite materials to carbonic acid and carbon dioxide (GFRP on the basis of UP or VE resins, depending on the operating conditions)

Table 68: Continued

Thermoplastic material	Medium	Concentration	Temperature K (°C)	Resistance	References
Tetrafluoroethylene-hexafluoropropylene copolymer (FEP)	CO_2 gas, moist	technically pure	293–423 (20–150)	+	[297]
	CO_2 gas, dry	technically pure	293–423 (20–150)	+	[297]
Ethylene chlorotrifluoroethylene copolymer (ECTFE)	CO_2 gas, moist	technically pure	293–393 (20–120)	+	[297]
	CO_2 gas, dry	technically pure	293–393 (20–120)	+	[297]
Ethylene tetrafluoroethylene copolymer (ETFE)	CO_2 gas, moist	technically pure	293–423 (20–150)	+	[297]
	CO_2 gas, dry	technically pure	293–393 (20–120)	+	[297]
Polyvinylidene fluoride (PVDF)	aqueous CO_2 solution	≤ saturated	≤ 373 (≤ 100)	+	[168]
	CO_2 gas, moist	technically pure	293–353 (20–80)	+	[297]
	CO_2 gas, dry	technically pure	293–373 (20–100)	+	[297]

+ = resistant ⊕ = moderately resistant

Table 68: Resistance of thermoplastic GFRP composite materials to carbonic acid and carbon dioxide (GFRP on the basis of UP or VE resins, depending on the operating conditions)

E
Material recommendations

This book has been compiled from literature data with the greatest possible care and attention. The statements made in this book only provide general descriptions and information.

Even for the correct selection of materials and correct processing, corrosive attack cannot be excluded in a corrosion system as it may be caused by previously unknown critical conditions and influencing factors or subsequently modified operating conditions.

No guarantee can be given for the chemical stability of the plant or equipment. Therefore, the given information and recommendations do not include any statements, from which warranty claims can be derived with respect to DECHEMA e.V. or its employees or the authors.

The DECHEMA e.V. is liable to the customer, irrespective of the legal grounds, for intentional or grossly negligent damage caused by their legal representatives or vicarious agents.

For a case of slight negligence, liability is limited to the infringement of essential contractual obligations (cardinal obligations). DECHEMA e.V. is not liable in the case of slight negligence for collateral damage or consequential damage as well as for damage that results from interruptions in the operations or delays which may arise from the deployment of this book.

A Metallic materials
A 2 Aluminium
A 3 Aluminium alloys

Aluminium and its alloys are resistant to dry carbon dioxide gas at ambient temperatures. In moist carbon dioxide gas and aqueous carbon dioxide-containing solutions these materials can be used if the pH value is not lower than 4 (Figure 8). However, the pH value drops below this level only at very high partial CO_2 pressures.

A 7 Copper
A 8 Copper-aluminium alloys
A 9 Copper-nickel alloys
A 10 Copper-tin alloys (bronze)

The corrosion behavior of copper in waters is determined by the extent to which cover layers are formed. The alkaline earth ions dissolved in water and the carbonate/hydrogen carbonate system and therefore the dissolved carbon dioxide play an important role in the formation of these cover layers. Under unfavorable conditions, there is a risk of pitting corrosion. This risk can be assessed with the help of the equations (page 38–39) described under A7:

$$Q = \frac{2 \times c \text{ (alkaline earth ions)}}{c \text{ (hydrogen carbonate ions)}} - 1 \text{ Mol/m}^3$$

There is a low risk of pitting corrosion if $Q < 0.15$ and $Q > 2$.
There is a higher risk of pitting corrosion if $Q = 0.5$ to 1.

$$S = \frac{c \text{ (hydrogen carbonate ions)}}{c \text{ (sulfate ions)}} \geq 2$$

at pH values > 7.5 the probability of pitting corrosion is low [55].

The copper-aluminium alloys are sufficiently resistant in carbon dioxide-containing waters and in moist gases.

Carbon dioxide dissolved in vapor systems or condensates may cause damage to copper-nickel materials under certain conditions.

Copper-tin alloys can be used in carbon dioxide-containing waters in a temperature range from 303 to 323 K (30°C to 50°C).

The good behavior in the atmosphere is not impaired by the contents of sulfur dioxide and carbon dioxide.

A 14 Unalloyed and low-alloy steels/cast steel
A 15 Unalloyed cast iron and low-alloy cast iron

Under pressure and at high temperatures carbon dioxide may cause local scale breakups in unalloyed and low-alloy steels, which may occur even at a very long time. This risk is increased with increasing temperature, increasing pressure, increasing water vapor content of the carbon dioxide and a higher surface roughness of the components. The low-alloy, temperature resisting, ferritic steels containing molybdenum, chromium or molybdenum and chromium can be used up to temperatures of about 723 K (450°C). At higher temperatures the increase of the oxidation rate is linear, rising with the increasing total pressure of the gas and with the increasing content of carbon dioxide in the gas.

CO_2 dissolved in aqueous media reacts to form carbonic acid, reducing the pH value of the solutions. This may cause a considerable corrosive attack on unalloyed and low-alloy steels. The corrosion rates to be expected as a function of the partial pressure of CO_2 and the temperature can be estimated by using Figure 27 and need to be taken into account by considering a corrosion allowance when specifying the wall thickness, unless other corrosion protection measures are possible. Improvements cannot be achieved by alloying elements of a magnitude present in low-alloy steels.

Higher strength steels, in particular heat treatable steels, may be damaged by stress corrosion cracking in moist carbon dioxide and generally in the $CO_2/CO/H_2O$ system. Here, the main influencing factors are the strength of the steel, the level of loading, the partial pressure of CO_2 and the temperature. The cracking tendency rises as the strength of steel, the loading level and the temperature increase.

A 17 Ferritic chrome steels with < 13 % Cr

In carbon dioxide at elevated pressures and temperatures the ferritic heat-resistant chromium molybdenum steels with chromium contents of 9 % and 12 % exhibit a similar accelerated oxidation as the unalloyed and low-alloy steels. Below temperatures of 823 K (550°C) local breakthroughs do not occur. Higher silicon contents up to 1 % in the steel exert a favorable influence on the formation of protecting oxide layers.

In aqueous solutions of carbon dioxide the corrosion rate of chromium-alloyed steels decreases if the chromium content increases and is assumed to be about 0.005 mm/a (0.20 mpy) in steel with 12.0 % Cr.

A 18 Ferritic chrome steels with ≥ 13 % Cr
A 19 High-alloy multiphase steels
A 19.1 Ferritic/perlitic-martensitic steels

The steels with higher chromium contents are largely resistant against material consuming corrosion in carbon dioxide-containing waters. However, in aqueous solutions containing also hydrogen sulfide apart from carbon dioxide there is a risk for chromium-alloyed steels to be damaged by stress corrosion cracking. By reducing the carbon content and adding the alloying elements nickel, molybdenum, copper and nitrogen, the corrosion behavior of a common 13 % chrome steel in CO_2 and H_2S containing media can be clearly improved regarding both its resistance to material consuming corrosion and to stress corrosion cracking.

A 20 Austenitic chromium-nickel steels
A 21 Austenitic CrNiMo(N) steels
A 22 Austenitic CrNiMoCu(N) steels

Stainless austenitic steels can be used in oxidizing or also carburizing gas atmospheres, such as air, steam, carbon dioxide or combustion gases, in a temperature range from 573 to 973 K (300°C to 700°C). The austenitic 18-8 nickel chromium steels exhibit a good resistance to oxidation in pure carbon dioxide in a temperature range up to about 803 K (530°C) – irrespective of their chemical composition to a major extent. However, at higher temperatures they are inferior to the steels with higher chromium and nickel contents. Therefore, the steels used under these conditions usually contain not less than 20 % chromium. Chloride and sulfate-containing ash deposits resulting from corrosion by hot combustion gases in power plants can intensify the attack.

The austenitic stainless Cr-Ni steels exhibit a generally good resistance in aqueous carbon dioxide containing solutions with corrosion rates below 0.01 mm/a (0.39 mpy).

A 26 Nickel
A 27 Nickel-chromium alloys
A 28 Nickel-chromium-iron alloys (without Mo)
A 29 Nickel-chromium-molybdenum alloys

In carburizing gases containing carbon dioxide, carbon monoxide or hydrocarbons, the behavior of Ni-Cr or Ni-Cr-Fe alloys differs. In carbon dioxide, for instance, not only damage to the material by carburization but also by oxidation must be expected. Concurrent carburization and oxidation impede the formation of a protecting oxide layer such that Ni-Cr and Ni-Cr-Fe alloys should only be used in this atmosphere up to 1273 K (1000°C). Silicon or aluminium-alloyed Ni-Cr-Fe materials can be used also at temperatures above 1273 K (1000°C).

The NiCrFeMoCu alloy 825 (Incoloy® 825), i.e. a tinanium-stabilized alloy, does not only exhibit a good corrosion resistance both under oxidizing and reducing conditions, but also has a good resistance to hot gases, e.g. combustion gases with sulfur dioxide and steam.

In aqueous carbon dioxide-containing solutions the passivatable NiCr, NiCrFe and NiCrMo alloys are practically resistant. Stress corrosion cracking of these materials is not even found in CO_2/H_2S-containing solutions.

A 37 Tantalum, niobium and their alloys
A 38 Titanium and titanium alloys
A 40 Zirconium and zirconium alloys

The highly corrosion resistant materials of groups A 37, A 38 and A 40 are resistant to aqueous carbon dioxide-containing solutions under all conditions.

B Nonmetallic inorganic materials
B 3 Carbon and graphite

Graphite impregnated with phenolic resin or polytetrafluoroethylene (PTFE), which therefore is pressure-tight, as used for the manufacture of equipment components, is resistant to aqueous carbon dioxide solutions or carbon dioxide gas (moist or dry) up to about 473 K (200°C), whereas graphite impregnated with polyvinylidene fluoride (PVDF) can be used at a temperature up to about 413 K (140°C).

Non-impregnated graphite has a minimum resistance to dry carbon dioxide gas up to about 903 K (630°C); a strong chemical degradation occurs at 1173 K (900°C).

B 4 Binders for building materials (e.g. concrete, mortar)

Concrete types with a low water/cements ratio and a high cement content exhibit a fairly low carbonatization rate.

B 5 Acid-resistant building materials

Polymer concrete with unsaturated polyester resins (UP resins) as a binding agent is resistant to aqueous carbon dioxide solutions at ambient temperature.

B 8 Enamel

Enamel coatings, if exposed to aqueous carbon dioxide solutions, have low corrosion rates of 0.1 mm/a (3.94 mpy) up to temperatures of 413 to 433 K (140 to 160°C) in a pH range of 4 to 6 and can be used up to about 453 K (180°C) if exposed to carbon dioxide gas.

B 12 Oxide ceramic materials

Ceramic materials of aluminium oxide, magnesium oxide, beryllium oxide or stabilized zirconium dioxide are resistant to carbon dioxide gas up to high temperatures (1473 K (1200°C)).

B 13 Metal-ceramic materials

Silicon carbide (SiC) produced by pressureless sintering is resistant to aqueous carbon dioxide solutions or moist carbon dioxide up to about 473 K (200°C). Compared to dry carbon dioxide, silicon carbide is resistant at 1273 K (1000°C), boron carbide at 1373 K (1100°C), molybdenum disilicide at 1873 K (1600°C) and cerium sulfide at 2273 K (2000°C).

C   Organic materials/Plastics

Thermoplastics

Thermoplastics widely used in the manufacture of chemical equipment, such as PE, PP and PVC-U, are suited up to a temperature range of 313 to 333 K (40 to 60°C) (PVC-U), up to 333 K (60°C) (PE) or up to 333–353 K (60–80°C) (PP) if exposed to aqueous carbon dioxide solutions or (moist or dry) carbon dioxide gas. The upper application temperature of PVC-C, which is also used in the manufacture of chemical equipment, is 333 to 353 K (60 to 80°C).

The fully fluorinated fluorothermoplastics PFA, MFA and FEP are resistant to aqueous carbon dioxide solutions or moist carbon dioxide gas up to 423 K (150°C), whereas PTFE is resistant up to about 473 K (200°C) and the partially fluorinated polymers can be used up to about 393 to 423 K (120 to 150°C) (ECTFE and ETFE) or up to 373 to 393 K (100 to 120°C) (PVDF). Regarding PVDF, the clearly higher water vapor permeability needs to be observed.

The other polymers mentioned in the Section C headed "Thermoplastics" are of subordinate importance as materials for the manufacture of chemical equipment and for components in process engineering plants. Their behavior in contact with carbon dioxide gas or aqueous carbon dioxide solutions corresponds to the durability specifications contained in section C.

Thermoplastic elastomers

The thermoplastic elastomers are resistant or widely resistant to carbon dioxide gas at ambient temperature.

Duroplastics

Duroplastics on the basis of EP resins are resistant to aqueous carbon dioxide solutions or moist carbon dioxide at ambient temperature. The resistance at elevated temperatures, e.g. at 323 to 333 K (50 to 60°C) strongly depends on the type of the hardener and the hardening conditions; warm and hot hardened EP resins exhibit a higher resistance compared to systems hardened at ambient temperature.

UP duroplastics on the basis of orthophthalic acid are resistant up to 303 to 313 K (30 to 40°C), those on the basis of iso- and terephthalic acid up to 333 to 343 K (60 to 70°C) and on the basis of HET acid or bisphenol A up to 343 to 353 K (70 to 80°C). For VE duroplastics a resistance up to about 353 K (80°C) can be assumed. The resistance to dry carbon dioxide gas is noticeably higher and can reach up to 473 K (200°C) in case of VE novolak resins.

The resistance of the polyurethanes strongly depends on the type of the polyol component; polyether polyols and polyols of dimerized fatty acids cause a higher resistance compared to polyester polyols. At room temperature polyurethanes are resistant or widely resistant to aqueous carbon dioxide solutions or (moist or dry) carbon dioxide gas.

Duroplastics are frequently used in combination with fillers or fibers. Since these components are often present in considerable concentration, they have an influence on the resistance of duroplastic materials.

Elastomers

For an exposure to aqueous carbon dioxide solutions or (moist or dry) carbon dioxide gas, in particular elastomers on the basis of EPDM, NBR, CSM and FKM are suited; these elastomers can also be used at an elevated temperature (up to 353 to 373 K (80 to 100°C)). FKM elastomers, however, are not suited for an exposure to carbon dioxide under pressure, since it involves the risk of explosive decompression. The highest resistance (up to about 503 K (230°C)) is provided by FFKM elastomers; however, their high price exerts a limiting effect on their use. Similarly to duroplastics, it should be noted that the resistance of elastomer materials is influenced by the vulcanization system, the fillers and the other additives.

D   Materials with special properties

The resistance of the organic coatings and linings, of the sealing materials and of the fiber laminates (GFRP and thermoplastic GFRP composite materials) are specified in the respective sections under D. In addition, Table 69 lists organic materials for components in process plants suitable for an exposure to aqueous carbon dioxide solutions or moist carbon dioxide. Usually, dry carbon dioxide is less aggressive, and hence it is not included in the table. In the column headed "Place of utilization" either utilization in the factory or utilization on the construction site is specified. Materials specified for utilization on the construction site can also be utilized in the factory, on condition that the components or equipment units are transportable. However, for the materials specified for utilization in the factory, certain devices and facilities are required there that are normally not available on the construction site.

E Material recommendations | 189

Equipment units / components in process engineering plants	Maximum application temperature K (°C)	Material	Thickness[1] mm	Place of utilization
Steel components: Storage containers, tanks, reactors, absorbers, evaporators, pipes[2], fittings	RT-313 (40)	thin coating, EP	0.3–0.4	construction site
	RT-313 (40)	thin coating, PF, hot-curing	0.2–0.3	factory
	RT	spray-applied coating, PUR	1–2	construction site
	303–333 (30–60)	spray-applied coating, EP	0.8–2	construction site
	303–323 (30–50)	spray-applied coating, UP[3], VE	0.8–1.5	construction site
	313–333 (40–60)	trowel-applied coating, UP[3], VE	2	construction site
	323–353 (50–80)	laminate coating, EP	3	construction site
	313–343 (40–70)	laminate coating, UP[3], VE	3	construction site
	403–423 (130–150)	powder coating, PFA/FEP	1–1.8	factory
	373–393 (100–120)	ECTFE, ETFE	0.5–2	factory
	353–373 (80–100)	hard rubber lining, NR, SBR, IR	4	factory
	353 (80)	hard rubber lining, NR, SBR, IR	4	construction site
	343–373 (70–100)	soft rubber lining, IIR, BIIR, CIIR	4	factory
	RT	soft rubber lining, CR	4	factory
	343–373 (70–100)	soft rubber lining, BIIR, cold vulcanized	4	construction site
	RT	soft rubber lining, CR, cold vulcanized	4	construction site
	333–353 (60–80)	soft rubber lining, CIIR, pre-vulcanized	4	construction site
	383 (110)	thermoplastic linings, glued, ECTFE, ETFE	2.3	factory
	423–473 (150–200)	loose jacket lining, PFTE	4–8	factory

[1] Layer thickness for coatings and linings
[2] For coatings partially limited pipe length and diameter
[3] UP resin on the basis of alcoxylated bisphenol A
RT = room temperature

Table 69: Maximum application temperatures of organic materials for components in process plants if exposed to aqueous carbon dioxide solutions or moist carbon dioxide (at normal pressure)

Table 69: Continued

Equipment units / components in process engineering plants	Maximum application temperature K (°C)	Material	Thickness[1] mm	Place of utilization
Plastic components: Storage containers, tanks, stirring vessels, reactors, absorbers, evaporators, pipes, fittings	313–333 (40–60)	PVC-U		factory
	333 (60)	PE		factory
	313–353 (60–80)	PP, PVC-C		factory
	373–393 (100–120)	PVDF, ECTFE, ETFE, PFA		factory
	RT–313 (40)	GFRP, UP		factory
	353 (80)	GFRP, VE		factory
	333–343 (60–70)	GFRP, EP, hot-curing		factory
	333 (60)	PVC-U/GFRP, PE/GFRP, PP/GFRP		factory
	353 (80)	PVC-C/GFRP		factory
	373–393 (100–120)	PVDF/GFRP, ECTFE/GFRP, ETFE/GFRP		factory
	423 (150)	FEP/GFRP, PFA/GFRP		factory
	423 (150)	PTFE (modified)/GFRP		factory
	373 (100)	furan resin/short carbon fibers/fillers, autoclave cured		factory
	373 (100)	phenolic resin/short carbon fibers/fillers, autoclave-cured		factory
Pumps and fittings – coatings	393–413 (120–140)	powder coating, fluorothermoplastics ECTFE, PFA/FEP	1–1.8	factory
– linings	353 (80)	hard rubber lining, NR, SBR, IR	4	factory
	333 (60)	soft rubber lining, IIR, BIIR	4	factory
	353–373 (80–100)	PE-UHMW, PVDF, PFA	2.3	factory
– secondary seals	373 (100)	EPDM, NBR		

[1] Layer thickness for coatings and linings
[2] For coatings partially limited pipe length and diameter
[3] UP resin on the basis of alcoxylated bisphenol A
RT = room temperature

Table 69: Maximum application temperatures of organic materials for components in process plants if exposed to aqueous carbon dioxide solutions or moist carbon dioxide (at normal pressure)

Table 69: Continued

Equipment units / components in process engineering plants	Maximum application temperature K (°C)	Material	Thickness[1] mm	Place of utilization
Seals				
– flat gaskets	443 (170)	synthetic fibers/elastomers, glass fibers/elastomers, carbon fibers/elastomers		
	523 (250)	PTFE without or with fillers		
	> 523 (> 250)	graphite expanded, without or with metal insert		
– packings	373–423 (100–150)	PTFE, carbon or synthetic yarns without or with PTFE or graphite impregnation		
– O rings	353–423 (80–150)	EPDM, NBR, FFKM		
Basins, shafts, channels made of concrete	313–353 (40–80)	laminate coatings, EP, UP [3], VE	3	construction site
	333 (60)	PE-HD, PP-H, PVC-U webs, mechanically anchored	3–5	construction site
	RT	soft rubber lining, CR, cold vulcanized	3	construction site
	343–353 (70–80)	soft rubber lining, BIIR, cold vulcanized	3–4	construction site
	333–353 (60–80)	soft rubber lining, CIIR, pre-vulcanized	3–4	construction site
Retention devices made of concrete for water polluting liquids	RT	self-leveling coating, EP	2–3	construction site
	RT	laminate coatings, EP, VE	2–3	construction site
	RT	PE-HD-, PP-H-, PVC-U-, PIB sheets, mechanically anchored, loose or glued	> 2	construction site
Floor coatings on concrete	RT-323 (50)	self-leveling, trowel-applied or stone floor coatings, EP, PUR, UP, VE	1–>10	construction site
	323–373 (50–100)	combined covering of ceramic tiles, EP, VE or FU putties and sealing layer	>20	construction site

[1] Layer thickness for coatings and linings
[2] For coatings partially limited pipe length and diameter
[3] UP resin on the basis of alcoxylated bisphenol A
RT = room temperature

Table 69: Maximum application temperatures of organic materials for components in process plants if exposed to aqueous carbon dioxide solutions or moist carbon dioxide (at normal pressure)

Bibliography

[1] Römpp Lexikon Chemie
Georg Thieme Verlag, Stuttgart/New York, 2005

[2] Topham, S.
Ullmann's Encyclopedia of Industrial Chemistry, 7 Edition, 2006
Electronic Release
Wiley-VCH Verlag, Weinheim,

[3] Firmenschrift
Schreckenberg, W.; Palmen, A.; Schubert, M.
Gase – Handbuch, Broschüre 90.1001, 2. Auflage
Messer Griesheim GmbH, Düsseldorf

[4] Hollemann, A. F.; Wiberg, E.
Lehrbuch der anorganischen Chemie

[5] Hahn, H.; Hempel, M.
Dauerschwingverhalten von Stahl in gashaltigen Korrosionslösungen
Technische Überwachung 13 (1972) 12, pp. 372–377

[6] Control of corrosion in cooling waters
in: Harston, J. D.; Ropital, F.
Control of corrosion in cooling waters, Bd. 40
European Federation of Corrosion, Brüssel, 2004

[7] Wortmann, G.
Die wasserseitige Korrosion in Wasser- und Dampfheizungsanlagen
Technische Überwachung 5 (1964) 8, pp. 292–296

[8] Tödt, F.
Korrosion und Korrosionsschutz
Walter de Gruyter & Co., Berlin, 1961

[9] Firmenschrift
Tabellen für das Labor
Fa. E. Merck, Darmstadt

[10] Desmaison, J.; Bouzovita, K.; Desmaison-Brit, M.; Billy, M.
Oxidation behaviour of transition metals of group IVa (titanium, hafnium) and Va (vanadium, tantalum) in carbon dioxide-carbon monoxide mixtures
8th European Congress of Corrosion
Vol. 1, 19–21
Nice, France, Nov. 1985,

[11] Grabke, H. J. [Ed.]; Schütze, M. [Ed.]
Corrosion by carbon and nitrogen; Metal dusting, carburisation and nitridation,
European Federation of Corrosion Publications Number 41, 2007

[12] Schmitt, G.
Der Korrosionsbegriff bei nichtmetallischen Werkstoffen
Materials and Corrosion 55 (2004) 5, pp. 367–372

[13] Carlowitz, B.
Kunststoff-Tabellen, 4. Aufl.
Carl Hanser Verlag, München, Wien, 1995

[14] Oberbach, K.; Baur, E.; Brinkmann, S.; Schmachtenberg, E.
Saechtling Kunststoff Taschenbuch, 29. Aufl.
Carl Hanser Verlag, München Wien, 2004

[15] Firmenschrift
Neoflon® FEP Pellets, EG-61m AK, 05/2003
Daikin Industries, Ltd., Osaka (Japan)

[16] Pauly, S.
Permeability and Diffusion Data
in: Brandrup, J.; Immergut, E. H.; Grulke, E. H.
Polymer Handbook, 4. Aufl., Bd. 6
John Wiley & Sons, New York, 1999, pp. 543–569

[17] Nakagawa, T.; Naruse, A.; Higuchi, A.
Permeation of Dissolved Carbon Dioxide in Synthetic Membranes
Journal of Applied Polymer Science 42 (1991) 2, pp. 383–389

[18] Marais, S.; Hirata, Y.; Langevin, D.; Chappey, C.; Nguyen, T. Q.; Metayer, M.
Permeation and Sorption of Water and Gases Through EVA Copolymers Films
Materials Research Innovations 6 (2002) 2, pp. 79–88

[19] Mohr, J. M.; Paul, D. R.
Comparison of Gas Permeation in Vinyl and Vinylidene Polymers
Journal of Applied Polymer Science 42 (1991), pp. 1711–1720

[20] Flaconneche, B.; Martin, J.; Klopffer, M. H.
Permeability, Diffusion and Solubility of Gases in Polyethylene, Polyamide 11 and Poly(vinylidene fluoride)
Oil & Gas Science and Technology 56 (2001) 3, pp. 261–278

[21] Villaluenga, J. P. G.; Seoane, B.
Influence of drawing on gas transport mechanism in LLDPE films
Polymer 39 (1998) 17, pp. 3955–3965

[22] Firmenschrift (CD-ROM)
Chemische Widerstandsfähigkeit, Simchem Version 5.0, 05/2003
Simona AG, Kirn

[23] Firmenschrift
Fluon® ETFE Ethylene Tetrafluoroethylene Copolymer, 01.12. 2001
Asahi Glass Compay, Ltd., Tokio (Japan)

[24] Firmenschrift
Solef® Polyvinylidene fluoride from Solvay – Technical Manual, B-1527 c-B-1-0901, 2001
Solvay S.A., Brüssel (Belgium)

[25] Bos, A.; Pünt, I. G. M.; Wessling, M.; Strathmann, H.
CO_2-induced plasticization phenomena in glassy polymers
Journal of Membrane Science 155 (1999) 1, pp. 67–78

[26] BASF AG, Ludwigshafen
Verhalten von Ultrason® gegen Chemikalien (Online in the Internet)
<http://www2.basf.de>
(downloaded 01.12.2007)

[27] Kapantaidakis, G. C.; Kaldis, S. P.; Dabou, X. S.; Sakellaropoulos, G. P.
Gas permeation through PFS-PI miscible blend membranes
Journal of Membrane Science 110 (1996) 2, pp. 239–247

[28] Marchese, J.; Garis, E.; Anson, M.; Ochoa, N. A.; Pagliero, C.
Gas sorption, permeation and separation of ABS copolymer membrane
Journal of Membrane Science 221 (2003) 1–2, pp. 185–197

[29] McGonigle, E.-A.; Liggat, J. J.; Pethrick, R. A.; Jenkins, S. D.; Daly, J. H.; Hayward, D.
Permeability of N_2, Ar, He, O_2 and CO_2 through biaxially oriented polyester films – dependance on free volume
Polymer 42 (2001) 6, pp. 2413–2426

[30] Aguilar-Vega, M.; Paul, D. R.
Gas Transport Properties of Poly(2,2,4,4-tetramethyl Cyclobutane Carbonate)
Journal of Polymer Science: Part B: Polymer Physics 31 (1993) 8, pp. 991–1004

[31] Firmenschrift
Verhalten von Ultramid®, Ultraform® und Ultradur® gegen Chemikalien, TI-KTE/AS-28 d 136707, 11/2005
BASF AG, Ludwigshafen

[32] Arkema, Inc., USA
Rilsan® PA 11: Created from a renewable source (Online in the Internet)
<http://www.arkema-inc.com>
(downloaded 18.10.2007)

[33] Gülmüs, S. A.; Yilmaz, L.
Effect of Temperature and Membrane Preparation Parameters on Gas Permeation Properties of Polymethacrylates
Journal of Polymer Science: Part B: Polymer Physics 45 (2007), pp. 3025–3033

[34] Chiou, J. S.; Paul, D. R.
Gas sorption and permeation in poly(ethyl methacrylate)
Journal of Membrane Science 45 (1989) 1–2, pp. 167–189

[35] Houde, A. Y.; Krishnakumar, B.; Charati, S. G.; Stern, S. A.
Permeability of Dense (Homogeneous) Cellulose Acetate Membranes to Methane, Carbon Dioxide, and Their Mixtures at Elevated Pressures
Journal of Applied Polymer Science 62 (1996) 13, pp. 2181–2192

[36] Firmenschrift
Thermoplastische Polyurethan-Elastomere. Elastollan® – Materialeigenschaften, Z/M, Fro 163–10-00, 10/2000
Elastogran GmbH, Lemförde

[37] Tremblay, P.; Savard, M. M.; Vermette, J.; Paquin, R.
Gas permeability, diffusivity and solubility of nitrogen, helium, methane, carbon dioxide and formaldehyde in dense polymeric membranes using a new on-line permeation apparatus
Journal of Membrane Science 282 (2006) 1/2, pp. 245–256

[38] Beck, K.; Kreiselmaier, R.; Peterseim, V.; Osen, E.
Permeation durch elastomere Dichtungswerkstoffe. Grundlagen – Werkstoffeigenschaften – Entwicklungstrends
KGK Kautschuk Gummi Kunststoffe 56 (2003) 12, pp. 657–660

[39] Leisenheimer, B.; Fritz, T.; Oellrich, L.
Untersuchungen zum Permeabilitätsverhalten von CO_2 in Elastomeren für die Autoklimatisierung
KI Luft- und Kältetechnik (2000) 12, pp. 580–585

[40] Dworak, D. P.; Lin, H.; Freeman, B. D.; Soucek, M. D.
Gas Permeability Analysis of Photo-Cured Cyclohexyl-Substituted Polysiloxane Films
Journal of Applied Polymer Science 102 (2006) 3, pp. 2343–2351

[41] DIN EN ISO 1043–1 (06/2002)
Kunststoffe – Kennbuchstaben und Kurzzeichen; Teil 1: Basis-Polymere und ihre besonderen Eigenschaften
Beuth Verlag GmbH, Berlin

[42] DIN ISO 1629 (11/2004)
Kautschuk und Latices – Einteilung, Kurzzeichen
Beuth Verlag GmbH, Berlin

[43] Rabalt, E.
Corrosion Guide, Elsevier Publishing Company,
Amsterdam, London, New York, 1968
Silber und Silberlegierungen, Ausgabe 1971
Degussa, Frankfurt/Main

[44] Aluminium Taschenbuch, 15. Aufl., Bd. 1
Aluminium Verlag, Düsseldorf, 1998, p. 330

[45] Bjorgum, A.; Sigurdson, H.; Nisanciuoglu, K.
Corrosion of commercially pure Al 99.5 in chloride solutions containing carbon dioxide, bicarbonate, and copper ions
Corrosion 51 (1995) 7, pp. 544–557

[46] Newton, jr. L. E.; Hausler, R. H.
CO_2 corrosion in oil and gas production
National Association of Corrosion Engineers, 1440 South Creek Drive, Houston, Texas 77084, USA, 1984, pp. 19–31

[47] Montrone, E. D.; Long, W. P.
Choosing materials for CO_2 absorption systems
Chemical Engineering 25 (1971) 1, pp. 94–99

[48] Reuter, M.
Schäden und Vorschriften bei Anlagen zur Förderung, Reinigung und zum Transport von Sauergas
Technische Überwachung 15 (1974) 3, pp. 65–70

[49] Winnacker, Küchler
Chemische Technologie, Band 4, Metalle, 1986

[50] Rickert, P.; Hampp, W.
Metalloberfläche 5 (1951) A, pp. 33–37

[51] Bay, J.
Löten von Kupferrohren für Reinst- und Medizinalgase
Technica 43 (1994) 4, pp. 66–68

[52] Wendler-Kalsch, E.
Korrosionsverhalten und Deckschichtbildung auf Kupfer und Kupferlegierungen
Zeitschrift für Werkstofftechnik (1982), pp. 129–137

[53] Kristiansen, H.
Corrosion of copper by water of various temperatures and carbon dioxide contents
Werkstoffe und Korrosion 28 (1977) 11, pp. 744–748

[54] Broo, A. E.; Beghult, B.; Hedberg, T.
Copper corrosion in drinking water distribution systems – The influence of water quality
Corrosion Science 36 (1997) 6,
pp. 1119–1132

[55] DIN EN 12502-2
(März 2005)
Korrosionsschutz metallischer Werkstoffe
Korrosionswahrscheinlichkeit in Wasserleitungssystemen
Teil 2: Übersicht über Einflussfaktoren für Kupfer und Kupferlegierungen
Beuth Verlag GmbH, 10772 Berlin

[56] Wendler-Kalsch, E.; Gräfen, H.
Korrosionsschadenkunde
Springer-Verlag, Berlin, 1988

[57] Wollrab, O.
Über den Einfluss der Wasserbeschaffenheit auf die Lochkorrosion in Trinkwasserleitungen aus Kupfer
Schadenprisma 18 (1989) 3, pp. 45–48

[58] von Franque, O.; Gerth, D.; Winkler, B.
Ergebnisse von Untersuchungen an Deckschichten in Kupferrohren
Werkstoffe und Korrosion 26 (1975) 4,
pp. 255–258

[59] Edwards, M.; Schock, M. R.; Meyer, T. E.
Alkalinity, pH, and copper corrosion by-product release
J. Am. Water Work Assoc. 88 (1996) 3,
pp. 81–94

[60] Monroe, E. S.
Effects of CO_2 in steam systems
Chemical Engineering 88 (1981) 6,
pp. 209–212

[61] Asrar, N.; Mali, A. U.; Ahmad, S.; et al.
Early failure of cupro-nickel condenser tubes in thermal desalination plant
Desalination 116 (1998), pp. 135–144

[62] Werkstoff-Datenblätter
CuSn10-C; CuSn7Zn4Pb7-C;
CuSn7Zn2Pb3-C; CuSn5Zn5Pb5-C;
CuSn5Pb1
Deutsches Kupferinstitut, Düsseldorf,
Stand 2005

[63] Rahmel, A.; Schwenk, W.
Korrosion und Korrosionsschutz von Stählen, 1. Aufl.
Verlag Chemie GmbH, Weinheim, 1977,
pp. 265–267

[64] Autorenkollektiv
Prüfung und Untersuchung der Korrosionsbeständigkeit von Stählen
Verlag Stahleisen, Düsseldorf, 1973

[65] Chengyu Yan; Oeters, F.
Kinetics of iron oxidation with CO_2 between 1300 and 1450 °C
Steel Research 65 (1994) 9, pp. 355–361

[66] Bredesen, R.; Kofstad, P.
On the oxidation of iron in CO_2 + CO mixtures: II.
Reaction Mechanisms During Initial Oxidation
Oxidation of Metals 35 (1991) 1–2,
pp. 107–137

[67] Bredesen, R.; Kofstad, P.
On the oxidation of iron in CO_2 + CO mixtures: III.
Coupled Linear Parabolic Kinetics
Oxidation of Metals 36 (1991) 1–2, pp. 27–56

[68] Schwenk, W.
Korrosion von unlegiertem Stahl in sauerstofffreier Kohlensäurelösung
Werkstoffe und Korrosion 25 (1974) 9,
pp. 643–646

[69] De Waard, C.; Milliams, D. E.
Carbonic acid corrosion of steel
Corrosion (Houston) 31 (1975) 5,
pp. 177–181

[70] Schmitt, G.; Rothmann, B.
Untersuchungen zum Korrosionsmechanismus von unlegiertem Stahl in sauerstofffreien Kohlensäurelösungen – Teil I. Kinetik der Wasserstoffabscheidung
Werkstoffe und Korrosion 28 (1977) 12,
pp. 816–822

[71] Schmitt, G.; Rothmann, B.
Untersuchungen zum Korrosionsmechanismus von unlegiertem Stahl in sauerstofffreien Kohlensäurelösungen – Teil II. Kinetik der Eisenauflösung
Werkstoffe und Korrosion 29 (1978) 2,
pp. 98–100

[72] Schmitt, G.; Rothmann, B.
Zum Korrosionsverhalten von unlegierten und niedriglegierten Stählen in Kohlensäurelösungen
Werkstoffe und Korrosion 29 (1978) 4,
pp. 237–245

[73] Schmitt, G.
NACE Corrosion 83
Paper 43
Houston, TX, USA, 1983

[74] McIntrire, G.; Lippert, J.; Yudelson, J.
The effect of dissolved CO_2 and O_2 on the corrosion of iron
Corrosion 46 (1990) 2, pp. 91–95

[75] Palacios, C. A.; Shadley, J. R.
Characteristics of corrosion scales on steel in CO_2 saturated NaCl brine
Corrosion 47 (1991) 2, pp. 122–127

[76] Jasinski, R.
Corrosion of N80-type steel by CO_2/water mixtures
Corrosion 43 (1987) 4, pp. 214–218

[77] Oblonsky, L. J.; Devine, T. M.
Corrosion of carbon steels in CO_2 saturated brine. A surface enhanced Raman spectroscopy study
Journal of electrochemical society 144 (1997) 4, pp. 1252–1260

[78] Cheng, Y. F.
Corrosion of X-65 pipeline steel in carbon dioxide-containing solutions
Bulletin of Electrochenistry 21 (2005) 11, pp. 503–511

[79] Schmitt, G.
Zur Frage der Korrosivität kohlensaurer Kondensate in Gasverteilungssystemen
GWF, Gas- Wasserfach: Gas/Erdgas 122 (1981) 2, pp. 49–54

[80] Carvalho, D. S.; Joia, C. J. B.; Mattos, O. R.
Corrosion rate of iron and iron-chromium alloys in CO_2 medium
Corrosion Science 47 (2005), pp. 2974–2986

[81] Okafor, P. C.; Nesic, S.
Effect of acetic acid on CO_2 corrosion of carbon steel in vapor-water two-phase horizontal flow
Chem. Eng. Comm. 194 (2007), pp. 141–157

[82] George, K. S.; Nesic, S.
Investigation of carbon dioxide corrosion of mild steel in the presence of acetic acid
Corrosion 63 (2007) 2, pp. 178–186

[83] Nesic, S.; Lunde, L.
Carbon dioxide corrosion of carbon steel in two-phase flow
Corrosion (Houston) 50 (1994) 9, pp. 717–727

[84] Nesic, S.; Solvi, G.T.; Enerhaug, J.
Comparison of the rotating cylinder and pipe flow tests for flow-sensitive carbon dioxide corrosion
Corrosion 51 (1995) 10, pp. 773–787

[85] Nesic, S.; Postlethwaite, J.; Olsen, S.
An electrochemical model for prediction of corrosion of mild steel in aqueous carbon dioxide solutions
Corrosion (Houston) 52 (1996) 4, pp. 280–294

[86] Amri, J.; Gulbrandsen, E.; Nogueira, R. P.
The effect of acetic acid on the pit propagation in CO_2 corrosion of carbon steel
Electrochemistry Communications 10 (2008), pp. 200–203

[87] Schmitt, G; Kunze, E.
Spannungsrißkorrosion niedriglegierter Vergütungsstähle in CO_2-haltigen wäßrigen Angriffsmitteln
in: Gräfen, H.; Rahmel, A.
Korrosion verstehen – Korrosionsschäden vermeiden, 1. Aufl., Bd. 1
Verlag Irene Kuron, Bonn, 1994, pp. 86–92

[88] Schmitt, G.
Untersuchungen zur Gefährdung hochfester Stähle durch wasserstoffinduzierte Spannungsrißkorrosion in Kohlendioxid enthaltenden Kondensaten
Werkstoffe und Korrosion 34 (1983) 4, pp. 187–198

[89] Gräfen, H.; Schlecker, H.
Spannungsrißkorrosion unlegierter und niedriglegierter Stähle in $CO/CO_2/H_2O/O_2$-haltigen Wässern
in: Gräfen, H.; Rahmel, A.
Korrosion verstehen – Korrosionsschäden vermeiden, 1. Aufl., Bd. 1
Verlag Irene Kuron, Bonn, 1994, pp. 93–97

[90] Yanateva, O. K.
Bulletin of the Academy of Science of the USSR, Division of Chemical Science (English Translation) (1954) pp.977–978

[91] Asahi, H.; Hara, T.; Sakamoto, S.
Corrosion properties and application limit of sour resistant 13 Cr steel tubing with improved CO_2 corrosion resistance
Conference: EUROCORR 97, Vol.1 Trondheim. Norway, 22–25 Sept. 1997

[92] Yamamoto, K.; Kagawa, N.
Ferritic stainless steels have improved resistance to SCC in chemical plant environments
Materials Performance 20 (1981) 6, pp. 32–37

[93] Lotz, U.; Schollmaier, M.; Heitz, E.
Flow-dependent corrosion. Ferrous materials in pure and particulate chloride solutions
Werkstoffe und Korrosion 36 (1985) 4, pp. 163–173

[94] Schollmaier, M.
Bericht zur Dechema-Jahrestagung 1984
4.3 Korrosions- und Verschleißerscheinungen in Zweiphasenströmungen
Werkstoffe und Korrosion 35 (1984), p. 429

[95] Lotz, U.; Schollmaier, M.; Heitz, E.
Flow-dependent corrosion– II. Ferrous materials in pure and particulate chloride solutions
Werkstoffe und Korrosion 36 (1985) 4, pp. 163–173

[96] Kohley, T.; Blatt, W.; Heitz, E.
Erosionskorrosion in Mehrphasensystemen der Offshore- und anderer Produktionstechniken – Teil C: Korrosionschemie und -schutzuntersuchungen unter Niederdruckbedingungen, unveröffentlichtes Manuskript
Forschungsstelle: Dechema-Institut, Frankfurt/Main

[97] Kimura, M.; Miyata, Y.; Yamane, Y.; Toyooka, T.; Nakano, Y.; Murase, F.
Corrosion resistance of high-strength modified 13 Cr steel
Corrosion (Houston) 55 (1999) 8, pp. 756–761

[98] Holm, R. A.; Evans, H. E.
The resistance of 20Cr/25Ni steels to carbon deposition. I. The role of surface grain size
Werkstoffe und Korrosion 38 (1987) 3, pp. 115–124

[99] Holm, R. A.; Evans, H. E.
The resistance of 20Cr/25Ni steels to carbon deposition. II. Internal oxidation and carburisation
Werkstoffe und Korrosion 38 (1987) 4, pp. 166–175

[100] Emsley, A. M.; Hill, M. P.
High-temperature oxidation of a 20/25 stainless steel in high-pressure carbon dioxide
Oxidation of Metals 33 (1990) 3/4, pp. 265–278

[101] Asher, J.; Sugden, S.; Benett, M. J.; Hawes, R. W. M.; Savage, D. J.; Price, J. B.
An investigation of the high temperature oxidation of a 20 Cr/25 Ni/Nb stainless steel in carbon dioxide using thin layer activation
Werkstoffe und Korrosion 38 (1987) 9, pp. 506–516

[102] Bennett, M. J.; Roberts, A. C.; Spindler, M. W.; Wells, D. H.
Interaction between oxidation and mechanical properties of 20Cr-25Ni-Nb stabilised stainless steel
Materials Science and Technology 6 (1990) 1, pp. 56–68

[103] Sroda, S.; Mäkipää, M.; Cha, S.; Spiegel, M.
The effect of ash deposition on corrosion behaviour of boiler steels in simulated combustion atmospheres containing carbon dioxide (CORBI PROJECT)
Materials and Corrosion 57 (2006) 2, pp. 176–181

[104] Montrone, E. D.; Long, W. P.
Choosing materials for CO_2 absorption systems
Chemical Engineering 78 (1971) 2, pp. 94, 96, 98–99

[105] Hamner, E.
Corrosion data survey, Fifth edition, 1974
NACE, Houston (Texas)

[106] Newton, L. E.; Hausler, R. H.
CO_2-Corrosion in oil and gas production
Selected papers, abstracts and references
1984 National Association of Corrosion Engineers, 1940 South Creek Drive, Houston, Texas, 77084

[107] Smith, A. F.
The duplex oxidation of vacuum annealed 316 stainless steel in CO_2/CO gas mixtures between 500 and 700 °C
Corrosion Science 24 (1984) 7, pp. 629–643

[108] Le Calvar, M.; Scott, M. P.; Magnin, T.; Rieux, P.
Strain oxidation cracking of austenitic stainless steels at 610 °C
Corrosion (Houston) 54 (1998) 2, pp. 101–105

[109] Lindström, R.; Svensson, J.-E.; Johansson, L.-G.
The influence of carbon dioxide on the atmospheric corrosion of some magnesium alloys in the presence of NaCl
Journal of the electrochemical society 149 (2002) 4, pp. B 103–B 107

[110] Brill, U.
Nickel, Cobalt und Nickel- und Cobalt-Basislegierungen
in: Egon Kunze
Korrosion und Korrosionsschutz, 1. Aufl., Bd. 2
Wiley-VCH, Berlin, 2001, pp. 1076–1168

[111] Hussain, N.; Schanz, G; Leistikow, S.; Shahid, K. A.
High-temperature oxidation and spalling behavior of Incoloy 825
Oxidation of Metals 32 (1989) 5–6, pp. 405–431

[112] Asphahani, A. I.
Evaluation of highly alloyed stainless materials for CO_2/H_2S environments
Corrosion 37 (1981) 6, pp. 327–335

[113] Falk, T.; Svensson, J.-E.; Johansson, L.-G.
The role of carbon dioxide in the atmospheric corrosion of zinc
Journal of the Electrochemical Society 145 (1998) 1, pp. 39–44

[114] Firmenschrift
Diabon® – Graphit für den Apparate- und Anlagenbau, 05/2006/1 E, 05/2006, und Mitteilung vom 30.11.2007
SGL Carbon GmbH, Meitingen

[115] GAB Neumann GmbH, Maulburg
Korrosionsbeständigkeit von imprägniertem Graphit (Online in the Internet)
<http://www.gab-neumann.de> (downloaded 20.04.2007)

[116] Firmenschrift
Chemische Beständigkeit von Kohlenstoff und Graphit, 08 00/05 NÄ, 08/2000
SGL Carbon GmbH, Bonn

[117] Lay, L. A.
Corrosion Resistance of Technical Ceramics, 2nd Ed.
HMSO Publications Centre, London (UK), 1991

[118] Firmenschrift
Chemische Beständigkeit von Grafit – Beständigkeit gegen Metalle, und Mitteilung vom 19.12.2003
SGL Carbon GmbH, Meitingen

[119] DePuy, G. W.
Chemical Resistance of Concrete
in: Klieger, P.; Lamond, J. F.
Significance of Tests and Properties of Concrete and Concrete-Making Materials
ASTM, Philadelphia (PA/USA), 1994, pp. 263–281

[120] Buenfeld, N. R.; Okundi, E.
Effect of cement content on transport in concrete
Magazine of Concrete Research 50 (1998) 4, pp. 339–351

[121] Sulapha, P.; Wong, S. F.; Wee, T. H.; Swaddiwudhipong, S.
Carbonation of Concrete Containing Mineral Admixtures
Journal of Materials in Civil Engineering 15 (2003) 2, pp. 134–143

[122] Claisse, P. A.; El-Sayad, H.; Shaaban, I. G.
Permeability and Pore Volume of Carbonated Concrete
ACI Materials Journal 96 (1999) 3, pp. 378–381

[123] Oye, B. A.; Justnes, H.
Carbonation Resistance of Polymer Cement Mortars (PCC)
in: Malhotra, V. M.
Durability of Concrete: 2. International Conference Montreal, Canada, Vol. 2
American Concrete Institute, Detroit (MI/USA), 1991, pp. 1031–1046

[124] Firmenschrift
Tabelle III: Chemische Beständigkeit von Polybeton; Mitteilung vom 16.09.2003
PRC Polymer-Kanalsystem GmbH & Co. KG, Rödinghausen

[125] Firmenschrift
Tycon® – Emaillierte Apparate für industrielle Prozesse, 94, 2003
Tycon Technoglass S.r.l., San Dona di Piave (Venezia/Italy)

[126] Firmenschrift
Glatter, reiner, resistenter. Das neue Pfaudler Pharmaglass PPG, 615–1d, 09/2000
Pfaudler Werke GmbH, Schwetzingen

[127] Firmenschrift
Pfaudler-Werkstoffe im Stahlverbund. Pfaudleremail WWG und Sonderemails, 206–5d 99.06.25, 06/1999
Pfaudler Werke GmbH, Schwetzingen

[128] Takayuki Narushima; Takashi Goto; Yoshio Yokoyama; Yasutaka Iguchi; Toshio Hirai
High-temperature active oxidation of chemically vapor-deposited silicon carbide in CO-CO_2 atmosphere
Journal of the American Ceramic Society, 76 (1993) 10, pp. 2521–2524

[129] Schlichting, J.
Siliciumcarbid als oxidationsbeständiger Hochtemperaturwerkstoff, Oxidations- und Heißkorrosionsverhalten I
Berichte Deutscher Keramischer Gesellschaft 56 (1979) 8, pp. 196–200

[130] Schlichting, J.
Siliciumcarbid als oxidationsbeständiger Hochtemperaturwerkstoff, Oxidations- und Heißkorrosionsverhalten II
Berichte Deutscher Keramischer Gesellschaft 56 (1979) 9, pp. 256–261

[131] Jacobsen, N. S.
Corrosion of silicon-based ceramics in combustion environments
Journal of the American Ceramic Society 76 (1993) 1, pp. 3–28

[132] Antill, J. E.; Warburton, J. B.
Active to passive transition in the oxidation of SiC
Corrosion Science 11 (1971), pp. 337–342

[133] Antill, J. E.; Warburton, J. B.
Oxidation of silicon and silicon carbide in gaseous atmospheres at 1000 to 1300 °C
in "Reactions between solids and gases"
AGARD Advisory Group for Aerospace Research and Development, Paris, France, 1970
Conference Proceedings No. 52 Paper No. 10

[134] Bremen, W.; Naoumidis, A.; Nickel, H.
Sliciumcarbid als Korrosionsschutzschicht auf graphitischen Werkstoffen für den HTR
Dissertation Techn. Hochschule Aachen, Jül-1422, 1977

[135] Bremen, W.; Naoumidis, A.; Nickel, H.
Oxidationsverhalten des pyrolytisch abgeschiedenen β-SiC unter einer Atmosphäre aus CO-CO_2-Gasgemischen
J. Nucl. Mat. 71 (1977), pp. 55–64

[136] Ebi, R.
Hochtemperaturoxidation von Siliciumcarbid und Siliciumnitrid in technischen Ofenatmosphären.
Dissertation Universität Karlsruhe 1973

[137] Elchin, V. I. et al.
Behaviour of a structural material with a high silicon carbide content in various gaseous media above 2000 K
Izvst. Akad. Nauk. SSSR, Neorgan. Mater. 7 (1971) 8, pp. 1342–1346

[138] Lee, K. N.; Jacobson, S.; Miller, R. A.
Refractory oxide coatings on SiC ceramics
MRS Bulletin XIX (1994) 10, S. 35–38

[139] Stott, F. H.; Mitchell, D. R. G.
The effects of high-temperature oxidation on the friction and wear of titanium-nitride-coated steel
Materials of High Temperatures 9 (1991) 4, pp. 185–192

[140] Mitchell, D. R. G.; Stott, F. H.
The oxidation of titanium nitride- and silicon nitride-coated stainless steel in carbon dioxide environments
Corrosion Science 33 (1992) 7, pp. 1083–1098

[141] Saint-Gobain Ceramics, Niagara Falls (NY/USA)
Hexoloy® Silicon Carbide Chemical Process Heat Exchanger Tubing (Online in the Internet)
<http://www.hexoloy.com>
(downloaded 24.10.2007)

[142] Chemical Resistance Volume I – Thermoplastics, 2nd Ed.
Plastics Design Library, Morris (NY/USA), 1994

[143] Borealis GmbH
Chemical resistance of polyolefins (Online in the Internet)
<http://www.borealisgroup.com>
(downloaded 20.04.2007)

[144] ISO/TR 10358 (06/1993)
Technical Report – Plastics pipes and fittings – Combined chemical-resistance classification table
Beuth Verlag GmbH, Berlin

[145] Herrlich; Land; Kunz; Michaeli
Kunststoffpraxis: Eigenschaften
WEKA Fachverlag, Augsburg, 2000

[146] Firmenschrift
Kunststoff-Armaturen, 03/2005
Frank GmbH, Mörfelden

[147] The Plastics Pipe Institute, USA
TR-19/2000 –Thermoplastics Piping for the Transport of Chemicals (Online in the Internet)
<http://spearsmfg.com>
(downloaded 21.04.2007)

[148] Firmenschrift
Trovidur® – Handbuch der chemischen Beständigkeit, 04.91/61870110/2.000, 04/1991
Haren Trovidur KG, Haren

[149] General Polymeric Corp., Reading (PA/USA)
Chemical Resistance of Porous Plastics (Online in the Internet)
<http://www.genpore.com>
(downloaded 20.10.2007)

[150] Firmenschrift
Planungsgrundlagen für industrielle Rohrleitungssysteme, 08/2006
Georg Fischer Piping Systems Ltd., Schaffhausen (Switzerland)

[151] Ineos Olefins & Polymers USA, League City, (TX/USA)
HDPE Chemical Resistance Guide (Online in the Internet)
<http://www.ineos-op.com>
(downloaded 04.11.2007)

[152] Iplex Pipelines, Sidney (Australia)
Chemical Resistance Chart (Online in the Internet)
<http://www.iplex.com.au>
(downloaded 29.11.2007)

[153] Beständigkeitsliste
in: Kunststoffrohrverband e. V. Bonn
Kunststoffrohr Handbuch – Rohrleitungssysteme für die Ver- und Entsorgung sowie weitere Anwendungsgebiete, 4. Aufl.
Vulkan-Verlag, Essen, 2000, p. 780

[154] Richtlinie DVS 2205–1 (04/2002)
Berechnung von Behältern und Apparaten aus Thermoplasten – Kennwerte
Verlag für Schweißen und verwandte Verfahren DVS-Verlag GmbH, Düsseldorf

[155] DIN 8075 Beiblatt 1 (02/1984)
Rohre aus Polyethylen hoher Dichte (HDPE). Chemische Widerstandsfähigkeit von Rohren und Rohrleitungsteilen
Beuth Verlag GmbH, Berlin

[156] ASV Stübbe GmbH & Co. KG, Vlotho
ASV – Beständigkeitstabelle (Online in the Internet)
<http://www.asv-stuebbe.de>
(downloaded 27.11.2007)

[157] Firmenschrift
GUR® PE-UHMW, GHR® PE-HMW – Beständigkeit gegen Chemikalien und andere Medien, B206 BRD-07.2001/042, 07/2001
Ticona GmbH, Kelsterbach

[158] Parker Hannifin Corp., USA
Industrial Hose Chemical Resistance Guide (Online in the Internet)
<http://www.safehose.com>
(downloaded 20.10.2007)

[159] Firmenschrift
marsoflex® – Beständigkeitsliste Chemieschläuche, 3000.3.03, 2003
Alfons Markert & Co. GmbH, Neumünster

[160] Semperit Technische Produkte GmbH, Wimpassing (Austria)
Chemie-Schläuche Beständigkeitsliste (Online in the Internet)
<http://www.semperflex.com>
(downloaded 29.11.2007)

[161] Firmenschrift
Isoplas® Crosslinkable Polyethylene, 15/09.98, 1998
Micropol Ltd., Stalybridge (England)

[162] DIN 8078 Beiblatt 1 (02/1982)
Rohre aus Polypropylen (PP). Chemische Widerstandsfähigkeit von Rohren und Rohrleitungsteilen
Beuth Verlag GmbH, Berlin

[163] Chemline Plastics Ltd., Thornhill (Ontario/Canada)
Chemical Resistance Guide (Online in the Internet)
<http://www.chemline.com>
(downloaded 29.11.2007)

[164] Ineos Olefins & Polymers USA, League City, (TX/USA)
PP Chemical Resistance Guide (Online in the Internet)
<http://www.ineos-op.com>
(downloaded 06.11.2007)

[165] M.K.Plastics Corp., Montreal (Quebec/Canada)
Corrosion Resistance Tables (Online in the Internet)
<http://www.mkplastics.com>
(downloaded 21.02.2008)

[166] Nibco, Inc., Elkhart (IN/USA)
Chemical Resistance Guide for Plastic and Metal Valves and Fittings (Online in the Internet)
<http://www.nibco.com>
(downloaded 29.11.2007)

[167] PVC Plastics Company, Inc., Evansville (IN/USA)
Chemical Resistance (Online in the Internet)
<http://www.e-pipeconnection.com>
(downloaded 30.11.2007)

[168] Medienlisten 40 (12/2000)
Medienlisten 40 für Behälter, Auffangvorrichtungen und Rohre aus Kunststoff
Deutsches Institut für Bautechnik, Berlin

[169] DIN 8061 Beiblatt 1 (02/1984)
Rohre aus weichmacherfreiem Polyvinylchlorid. Chemische Widerstandsfähigkeit von Rohren und Rohrleitungsteilen aus PVC-U
Beuth Verlag GmbH, Berlin

[170] DIN 8080 Beiblatt 1 (08/2000)
Rohre aus chloriertem Polyvinylchlorid (PVC-C) PVC-C 250. Allgemeine Güteanforderungen, Prüfung. Chemische Widerstandsfähigkeit
Beuth Verlag GmbH, Berlin

[171] P. E. P., Branchburg (NJ/USA)
CPVC Chemical Resistance (Online in the Internet)
<http://www.pep-plastic.com>
(downloaded 23.10.2007)

[172] Firmenschrift
Corzan® Industrial Systems – Chemical Resistance Data, 1201 CZ-1, 2001
Noveon Europe B. V. B. A., Brüssel (Belgium)

[173] Chemical Resistance Volume II – Thermoplastic Elastomers, Thermosets, and Rubbers, 2nd Ed.
Plastics Design Library, Norwich (NY/USA), 1996

[174] Kumazawa, H.; Wang, J.-S.; Naito, K.; Messaoudi, B.; Sada, E.
Gas Transport in Polymer Membrane at Temperatures Above and Below Glass Transition Point
Journal of Applied Polymer Science 51 (1994) 6, pp. 1015–1020

[175] Pal, S. N.; Ramani, A. V.; Subramanian, N.
Gas Permeability Studies on Poly(Vinyl Chloride) Based Polymer Blends Intended for Medical Applications
Journal of Applied Polymer Science 46 (1992) 6, pp. 981–990

[176] Sunny, M. C.; Ramesh, P.; George, K. E.
Effect of Partial Replacement of Di(2-ethyl hexyl)phthalate, by a Polymeric Plasticizer, on the Permeability and Leaching Properties of Poly(vinyl chloride)
Journal of Applied Polymer Science 102 (2006) 5, pp. 4720–4727

[177] Tapflo AB, Kungälv (Sweden)
Chemical Guide (Online in the Internet)
<http://www.tapflo.com>
(downloaded 24.10.2007)

[178] Val-Matic Corp., Elmhurst (IL/USA)
Val-Matic Chemical Resistance Guide for Valves (Online in the Internet)
<http://www.valmatic.com>
(downloaded 06.12.2007)

[179] DuPont, USA
High Performance Film: DuPont PFA fluorocarbon film (Online in the Internet)
<http://plastics.dupont.com>
(downloaded 16.10.2007)

[180] Symalit AG, Lenzburg (Switzerland)
Symalit Chemical Resistance (Online in the Internet)
<http://epp.quadrantplastics.com>
(downloaded 19.11.2007)

[181] DuPont, USA
High Performance Film: DuPont FEP fluorocarbon film (Online in the Internet)
<http://plastics.dupont.com>
(downloaded 17.10.2007)

[182] Firmenschrift
Chemical Resistance of Halar® Fluoropolymer
Ausimont USA, Inc., Morristown (NJ/USA)

[183] DuPont
DuPont Tefzel® fluoropolymer resin – Chemical Use Temperature Guide (Online in the Internet)
<http://www2.dupont.com>
(downloaded 21.12.2006)

[184] DVS Richtlinie 2205–1 Beiblatt 20 (08/2007)
Berechnung von Behältern und Apparaten aus Thermoplasten – PVDF-Medienliste
Verlag für Schweißen und verwandte Verfahren DVS-Verlag GmbH, Düsseldorf

[185] Firmenschrift
Hylar® Polyvinylidenefluoride – Chemical Resistance Chart, 0022
Ausimont USA, Inc., Morristown (NJ/USA)

[186] Firmenschrift
Solef® Chemical Resistance Table Inorganic Media, and information from 29.11.2004
Solvay Solexis, Brüssel (Belgium)

[187] Chevron Phillips Chemical Company LLC, The Woodlands (TX/USA)
Ryton® PPS Chemical Compatibility Chart (Online in the Internet)
<http://www.cpchem.com>
(downloaded 01.12.2007)

[188] Quadrant Engineering Plastic Products, Lenzburg (Switzerland)
Chemical Resistance Data (Online in the Internet)
<http://www.quadrantepp.com>
(downloaded 20.10.2007)

[189] Firmenschrift
Beständigkeit für Arlon®
Greene, Tweed u. Co. GmbH, Hofheim

[190] Firmenschrift
victrex® High Performance PEEK Polymers – Chemikalienbeständigkeit, 404/1.5m, 2004
Victrex Europa GmbH, Hofheim

[191] Solvay Solexis S. p. A., Bollate (Italy)
Hyflon® PFA Perfluoroalkoxy Fluorocarbon Resins – Design and Processing Guide (Online in the Internet)
<http://www.solvaysolexis.com>
(downloaded 30.11.2007)

[192] BASF AG, Ludwigshafen
Verhalten von Polystyrol gegen Chemikalien (Online in the Internet)
<http://www2.basf.de>
(downloaded 12.10.2007)

[193] BASF AG, Ludwigshafen
Beständigkeit von Styrolcopolymeren gegen Chemikalien (Online in the Internet)
<http://www2.basf.de>
(downloaded 12.10.2007)

[194] DuPont
Crastin® PBT und Rynite® PET – Thermoplastische Polyester – Konstruktionshinweise (Online in the Internet)
<http://www.plastics.dupont.com>
(downloaded 16.10.2007)

[195] DSM, Heerlen (Netherlands)
Resistance of DSM Engineering Plastics to Chemicals (Online in the Internet)
<http://www.dsm.com>
(downloaded 30.11.2007)

[196] SPS International, Woodland (CA/USA)
Chemical Resistance of Polycarbonate Sheets (Online in the Internet)
<http://www.palramhort.com>
(downloaded 20.10.2007)

[197] Rehau AG + Co., Rehau
RAU-PA Polyamide – Chemical resistance of Series RAU-PA 100 (polyamide 6) and RAU-PA 200 (polyamide 6.6) (Online in the Internet)
<http://www.rehau.de>
(downloaded 23.10.2007)

[198] Firmenschrift
Materialdaten für die Konstruktion mit Minlon® – Zytel® Polyamiden, H-56843-2, 02/1998
DuPont de Nemours International S. A., Genf (Switzerland)

[199] DuPont
Delrin® Acetalhomopolymere – Technische Informationen (Online in the Internet)
<http://www.plastics.dupont.com>
(downloaded 16.10.2007)

[200] Al-Enezi, S.; Hellgardt, K.; Stapley, A. G. F.
Mechanical Measurement of the Plasticization of Polymers by High-Pressure Carbon Dioxide
International Journal of Polymer Anal. Charact. 12 (2007), pp. 171–183

[201] Puleo, A. C.; Muruganandam, N.; Paul, D. R.
Gas Sorption and Transport in Substituted Polystyrenes
Journal of Polymer Science, Part B: Polymer Physics 27 (1989) 11, pp. 2385–2406

[202] Fu, Y.-J.; Hu, C.-C.; Lee, K.-R.; Tsai, H.-A.; Ruaan, R.-C.; Lai, J.-Y.
The correlation between free volume and gas separation properties in high molecular weight poly(methyl methacrylate) membranes
European Polymer Journal 43 (2007) 3, pp. 959–967

[203] Jimenez, A.; Thompson, G. L.; Matthews, M. A.; Davis, T. A.; Crocker, K; Lyons, J. S.; Trapotsis, A.
Compatibility of medical-grade polymers with dense CO_2
Journal of Supercritical Fluids 42 (2007), pp. 366–372

[204] Shieh, Y.-T.; Su, J.-H.; Manivannan, G.; Lee, P. H. C.; Sawan, S. P.; Spall, W. D.
Interaction of Supercritical Carbon Dioxide with Polymers. II. Amorphous Polymers
Journal of Applied Polymer Science 59 (1996) 4, pp. 707–717

[205] Khan, F.; Czechura, K.; Sundararajan, P. R.
Modification of morphology of polycarbonate/thermoplastic elastomer blends by supercritical CO_2
European Polymer Journal 42 (2006) 10, pp. 2899–2904

[206] Firmenschrift
Hytrel® Thermoplastic Polyester Elastomer. Fluid and chemical resistance guide, 06.99
L-11964, 06/1999
DuPont de Nemours, Wilmington (DE/USA)

[207] Firmenschrift
Chemikalienbeständigkeit von Desmopan®/TPU, und Mitteilung vom 10.11.2004
Bayer MaterialScience AG, Leverkusen

[208] Matsunaga, K.; Sato, K.; Tajima, M.; Yoshida, Y.
Gas Permeability of Thermoplastic Polyurethane Elastomers
Polymer Journal (Tokyo, Japan) 37 (2005) 6, pp. 413–417

[209] Zrunek Gummiwaren GmbH, Wien (Austria)
Elastomer Beständigkeiten (Online in the Internet)
<http://www.zrunek.at>
(downloaded 06.12.2007)

[210] DuPont Elastomers, Wilmington (DE/USA)
General Chemical Resistance Guide (Online in the Internet)
<http://dupontelastomers.com>
(downloaded 06.12.2007)

[211] Perlast Ltd., Blackburn (England)
A Technical Guide to Elastomer Compounds and Chemical Compatibility (Online in the Internet)
<http://www.prepol.com>
(downloaded 20.10.2007)

[212] Firmenschrift
Beständigkeitsliste für Elastomere, 2003
Artemis Kautschuk- und Kunststofftechnik, Hannover

[213] Firmenschrift
Beständigkeitsliste Elastomer- und PTFE-Werkstoffe, 1997
Busak + Shamban GmbH & Co. KG, Stuttgart

[214] Firmenschrift
Materials Compatibility Guide, 05/02, 05/2002
PCM Pompes, Vanves (France)

[215] Masterflex AG, Gelsenkirchen
Chemische Beständigkeitsliste K (Online in the Internet)
<http://www.masterflex.de>
(downloaded 06.12.2007)

[216] DuPont Elastomers, Wilmington (DE/USA)
Kalrez® Perfluoroelastomer Parts. Chemical Resistance and Fluid Compatibility, Including All Chemicals Under the Clean Air Act (Online in the Internet)
<http://www.dupontelastomers.com>
(downloaded 10.11.2007)

[217] Firmenschrift
Beständigkeit von Vulkollan® in verschiedenen Medien, PU 52672, 06/1982
Bayer MaterialScience AG, Leverkusen

[218] Härtel, G.; Rompf, F.
Dicht oder nicht dicht? Dichtungen quellen unter dem Einfluss von Kohlendioxid
Chemie Technik 25 (1996) 10, pp. 42–43

[219] George, A. F.; Sully, S.; Davies, O. M.
Carbon dioxide saturated elastomers: the loss of tensile properties and the effects of temperature rise and pressure cycling
BHR Group Conference Series Publication (1997) 26 (Fluid Sealing), pp. 437–457

[220] Jones, P. B.
Polymer emulsion based systems for the protection and repair of concrete
Journal of the Oil and Colour Chemists' Association 71 (1989) 12, pp. 407–409

[221] Kim, J.-W.; Lee, S.-M.; Hong, J.-K.; Lim, J.-C.; Kim, B.-S.; Park, S.; Hong, S. M.; Lee, H. K.; Park, J. M.
Water Vapor and CO_2 Permeabilities of Acrylic Latex Coatings
J. Ind. Eng. Chem. 7 (2001) 6, pp. 380–388

[222] Firmenschrift
Kunststoff-Beschichtungen im Produkte-Segment PE, EVA, PP, 12/2006
Epsosint AG, Pfyn (Switzerland)
[223] Firmenschrift
Epover®-S – Glas-Splitterschutz-Beschichtungen, 03/2004
Eposint AG, Pfyn (Switzerland)
[224] Plascoat Systems Ltd., Farnham (Surrey/UK)
Plascoat® PPA 571 – Performance Polymer Alloy Coating (Online in the Internet)
<http://www.plascoat.com>
(downloaded 04.11.2007)
[225] Firmenschrift
Rilsan® PA 11 – Korrosionsschutz / Verschleißschutz / Elektroisolation, 08/2003
Eposint AG, Pfyn (Switzerland)
[226] Firmenschrift
ECTFE Halar® – Thermoplastischer High-tech-Beschichtungs-Kunststoff, 05/2005
Eposint AG, Pfyn (Switzerland)
[227] Firmenschrift
Beständigkeitsliste Verfahren Proco-E-CTFE (Halar®), 03/1996
Hüni + Co. KG, Friedrichshafen
[228] Firmenschrift
Fluon® ETFE Lining – Fluon® ETFE Rotomolding, N.D.A.02.3, 2003
Asahi Glass Co., Ltd., Tokyo (Japan)
[229] Firmenschrift
Edlon® PFA Beständigkeitsliste, 01/05.07, 05/1997
Rudolf Gutbrod GmbH, Dettingen/Erms
[230] Firmenschrift
Jumbo – Der Standard für Fluor-Polymer-Beschichtungen und Mitteilung v. 12.04.2001
Rhenotherm Kunststoffbeschichtungs GmbH, Kempen
[231] Coatings and Linings for Immersion Service, 2nd Ed.
NACE International, Houston (TX/USA), 1998, pp. 67, 85, 169
[232] Firmenschrift
Proco-Kunststoffbeschichtungen®: Proco-L (F21), 05/2004, und Beständigkeitsliste Proco-Kunststoff-Beschichtungen, 08/2002
Hüni + CO KG, Friedrichshafen

[233] Atlas Minerals & Chemicals, Inc., Merztown (PA/USA)
Chemical Construction Materials (Online in the Internet)
<http://www.atlasmin.com>
(downloaded 07.12.2007)
[234] Firmenschrift (CD-ROM)
GBT – Technische Informationen – Beständigkeitstabellen, 06/2006
GBT-Bücolit GmbH, Marl
[235] Firmenschrift
Proguard® CN 100 – Resistenzliste und Produktdatenblatt vom 31.07.2002, und Mitteilung vom 28.11.2002
Ceram-Kote Europe GmbH, Rödinghausen
[236] Firmenschrift
Plasite Lining Systems – Produktdatenblattordner und CD-ROM Chemical Resistance Guides, 03/2000
Plasite Protective Coatings, Maple Shade, NJ/USA
[237] DIN EN 14879–2 (02/2007)
Beschichtungen und Auskleidungen aus organischen Werkstoffen zum Schutz von industriellen Anlagen gegen Korrosion durch aggressive Medien. Teil 2: Beschichtungen für Bauteile aus metallischen Werkstoffen; Deutsche Fassung EN 14879-2:2006
Beuth Verlag GmbH, Berlin
[238] Firmenschrift
Oxydur® UP 410, -VE-L, -VE-S, -Flake – Chemische Beständigkeiten, 210A, 05/1999
Steuler Industriewerke GmbH, Höhr-Grenzhausen
[239] Firmenschrift
Beständigkeitsliste Bodenbeschichtungen und Kitte, O-08/94-329, 08/1994
KCH Group GmbH, Siershahn
[240] Firmenschrift
Chemical Resistance Chart – Organic Monolithic Products – Mortars & Inorganic Monolithic Products, 02/2006
Sauereisen, Pittsburgh (PA/USA)
[241] Firmenschrift
Furadur®-Kitt, -F-Kitt – Chemische Beständigkeiten, 320, 07/1998
Steuler Industriewerke GmbH, Höhr-Grenzhausen

[242] Sybra Permatex International GmbH, Schweinfurt
Asplit® – Chemische Beständigkeit (Online in the Internet)
<http://www.sybra-int.com>
(downloaded 10.12.2007)

[243] AGI – Arbeitsblatt S 10 Teil 3 (09/2001)
Schutz von Baukonstruktionen mit Plattenbelägen gegen chemische Angriffe (Säureschutzbau); Plattenlagen
Callwey Verlag Leser-Service, Lindau

[244] DIN 28052–5 (04/1997)
Chemischer Apparatebau. Oberflächenschutz mit nichtmetallischen Werkstoffen für Bauteile aus Beton in verfahrenstechnischen Anlagen. Teil 5: Kombinierte Beläge
Beuth Verlag GmbH, Berlin

[245] Firmenschrift
Oxydur VE-K, TM, F, A – Chemische Beständigkeiten, 300, 10/1999
Steuler Industriewerke GmbH, Höhr-Grenzhausen

[246] Firmenschrift
Alkadur® K 75, Alkadur® SK, 330, 04/1999
Steuler Industriewerke GmbH, Höhr-Grenzhausen

[247] AGI Arbeitsblatt S 10–2 (11/2002)
Schutz von Baukonstruktionen mit Plattenbelägen gegen chemische Angriffe (Säureschutzbau) – Dichtschichten
Callwey Verlag Leser-Service, Lindau

[248] Firmenschrift
Oxydur® HT, -HTE, Oxydur® UP 82, Oxydur® UP 82 BW u. a. – Chemische Beständigkeiten, 200 und 200A, 12/1999
Steuler Industriewerke GmbH, Höhr-Grenzhausen

[249] FDT FlachdachTechnologie GmbH & Co. KG, Mannheim
Chemische Beständigkeit der Säureschutzbahnen Rhepanol® O.R.G. und Rhepanol® O.R.F. (Online in the Internet)
<http://www.fdt.de>
(downloaded 10.12.2007)

[250] Firmenschrift
Alkadur® D, DF, -DFG-LF, K47 F KB, -K 47 NW, -K47 F – Chemische Beständigkeiten, 230, 10/1999
Steuler Industriewerke GmbH, Höhr-Grenzhausen

[251] Firmenschrift (CD-ROM)
Boden- und Gewässerschutzsysteme – Beständigkeiten, Zulassungen, Broschüren, 02/2008
StoCretec GmbH, Kriftel

[252] Firmenschrift
Produktsysteme 2004/2005, Betoninstandsetzung und Oberflächenschutz, Industrieböden und WHG-Beschichtungen, Injektionssysteme, Fugenabdichtung, 2004
MC-Bauchemie Müller GmbH & Co. KG, Bottrop

[253] Firmenschrift
Beständigkeitsliste für Beschichtungssysteme mit allgemeinen bauaufsichtlichen Zulassungen des DIBt, SP-116.1/12-2002d, 2002
KCH Group GmbH, Siershahn

[254] Firmenschrift
Technische Dokumentation Bau, TMB/C/04–05/15.000, 04/2005
Sika Deutschland GmbH, Stuttgart

[255] Firmenschrift
Disbon Bautenschutz Werkstoff-Programm 2005, Le DG Ch 10 03/05, 2005
Caparol Farben Lacke Bautenschutz GmbH, Ober-Ramstadt

[256] Allgemeine bauaufsichtliche Zulassung Z-59.16-268
Beschichtungssystem "Alkadur HR" (Antragsteller: Steuler Industriewerke GmbH) vom 11.11.2002
Deutsches Institut für Bautechnik, Berlin

[257] Allgemeine bauaufsichtliche Zulassung Z-59.12-309
Beschichtungssystem "StoCretec WHG System 1" (Antragsteller: StoCretec GmbH) vom 02.08.2006
Deutsches Institut für Bautechnik, Berlin

[258] Firmenschrift
Chemical Resistance Chart and information from 20.12.2006
Polycorp Ltd., Elora (Ontario/Canada)

[259] Firmenschrift
Beständigkeitsliste Werksgummierungen, SP-116.2/01–2003def, 01/2003
SGL Acotec GmbH, Siershahn

[260] Electro Chemical Engineering & Manufacturing Co., Emmaus (PA/USA)
Corrosion Protection (Online in the Internet)
<http://www.electrochemical.net>
(downloaded 18.05.2007)

[261] Khaladkar, P. R.
Using Plastics, Elastomers, and Composites for Corrosion Control
in: Revie, R. W.
Uhlig's Corrosion Handbook, 2nd Ed.
John Wiley & Sons, Inc., New York, 2000, p. 989

[262] Firmenschrift (CD-ROM)
WAGU Produktinformation, 02/2007
WAGU Gummitechnik GmbH, Warstein

[263] Firmenschrift (CD-ROM)
Corrosion Protection Systems – Product Information, 05/2006
Tip Top Oberflächenschutz Elbe GmbH, Wittenberg

[264] Firmenschrift
Beständigkeitsliste Baustellengummierungen, SP-116.3/02-2003def, 02/2003
SGL Acotec GmbH, Siershahn

[265] Hölter, D.; Burkhart, T.
Spritzbare Hartgummi-Beschichtungen
KGK Kautschuk Gummi Kunststoffe 56 (2003) 12, pp. 646–649

[266] Blair Rubber Co., Akron (OH/USA)
Chemical Resistance Table (Online in the Internet)
<http//www.blairrubber.com>
(downloaded 10.12.2007)

[267] Firmenschrift
Trovidur® W 2000 – Auskleidungsfolie aus PVC-P – extrudiert –, 01.89/1.200.170 ha 0040-64, 01/1989
Röchling Engineering Plastics KG, Haren

[268] DIN 28091 Teil 1–4 (09/1995)
Technische Lieferbedingungen für Dichtungsplatten; Teil 1: Dichtungswerkstoffe; Allgemeine Festlegungen; Teil 2: Dichtungswerkstoffe auf Basis von Fasern (FA); Anforderungen und Prüfung; Teil 3: Dichtungswerkstoffe auf Basis von PTFE (TF); Anforderung und Prüfung; Teil 4: Dichtungswerkstoffe auf Basis von expandiertem Graphit (GR); Anforderungen und Prüfung
Beuth Verlag GmbH, Berlin

[269] Firmenschrift (CD-ROM)
NovaDisc – Die Software zur Berechnung dichtungstechnischer Rahmenbedingungen, Version 6.0, 02/2008
Frenzelit-Werke GmbH & Co. KG, Bad Berneck

[270] Firmenschrift
Gylon® Flachdichtungen, GP-D4.2.11/02, 11/2002
Beständigkeitsliste für Flachdichtungen, GP-D7.1.05/04, 05/2004
Garlock GmbH, Neuss

[271] DIN 28090 Teil 2 und 3
Statische Dichtungen für Flanschverbindungen. Teil 2 (09/1995): Dichtungen aus Dichtungsplatten; Spezielle Prüfverfahren zur Qualitätssicherung; Teil 3 (09/1995): Dichtungen aus Dichtungsplatten; Prüfverfahren zur Ermittlung der chemischen Beständigkeit
Beuth Verlag GmbH, Berlin

[272] Klinger GmbH, Idstein
Beständigkeitstabelle (Online in the Internet)
<http://www.klinger.de>
(downloaded 09.10.2007)

[273] Firmenschrift
AFM 34 – Beständigkeit gegenüber chemischen Medien, 04/04-39-00025-00, 04/2004
Reinz-Dichtungs-GmbH, Neu-Ulm

[274] W. L. Gore & Associates GmbH, Putzbrunn
Gore-Tex® Serie 300 und 600 Dichtungsband (Online in the Internet)
<http://www.gore.com/sealants>
(downloaded 18.05.2007)

[275] Firmenschrift
Stopfbuchspackungen, 2003
Freudenberg Process Seals KG, Viernheim

[276] Garlock GmbH, Neuss
Stopfbuchspackungen (Online in the Internet)
<http://www.garlock.eu.com>
(downloaded 10.11.2007)

[277] Firmenschrift
Stopfbuchspackungen zur Abdichtung von Spindeln, Wellen und Plungern in Armaturen, Pumpen, Rührwerken, Ventilatoren u. a., P6D/D2/5.000/08.03/1.2.1, 08/2003
Burgmann Dichtungswerke GmbH & Co. KG, Wolfratshausen

[278] Firmenschrift
Parker Seals – Medien Beständigkeits-Tabelle, 5703 G, 09/2000
Parker Hannifin GmbH, Pleidelsheim

[279] Firmenschrift
Simrit® Katalog, 30 D004 30.00405 ISRe, 2005
Freudenberg Simrit KG, Weinheim

[280] Firmenschrift
O-Ring Werkstoffe, 99D/013/011/0401, 04/2001
Busak + Shamban GmbH, Stuttgart

[281] Firmenschrift
Chemical Compatibility Guide Elastomeric Compounds, 03/06-OU-5M
GH-US-GE-015, 03/2006
Greene, Tweed & Co. GmbH, Hofheim am Taunus

[282] Trelleborg Sealing Solutions Germany
Isolast® Perfluorelastomer Dichtungen (Online in the Internet)
<http://www.tss.trelleborg.com>
(downloaded 29.02.2008)

[283] Firmenschrift
Burgmann Gleitringdichtungen Konstruktionsmappe 15.3, KM 15D/D5/8.000/10.05/ 1.2.1, 10/2005
Burgmann Industries GmbH & Co. KG, Wolfratshausen

[284] DSM Composite Resins
Guide to chemical resistance. Focusing on corrosion solutions (Online in the Internet)
<http://www.dsm.com>
(downloaded 10.12.2007)

[285] Firmenschrift
Hetron® and Aropol® Resin Selection Guide. For Corrosion Resistant FRP Applications, AC5–1102-C 00410MO, 2000
Ashland Inc., Dublin (OH/USA)

[286] Ashland Composite Polymers, Columbus (OH/USA)
Derakane® Epoxy Vinyl Ester Resins – Chemical Resistance Guide (Online in the Internet)
<http://www.derakane.com>
(downloaded 10.12. 2007)

[287] Nil-Cor LLC., Alliance (OH/USA)
Nil-Cor® Advanced Composite Valves – Chemical Resistance Guide (Online in the Internet)
<http://www.nilcor.com>
(downloaded 10.12.2007)

[288] Advanced Valve Technologies, Broadstairs (Kent/England)
Advanced Valve Technologies Binder 2003 (Online in the Internet)
<http://www.advalve.com>
(downloaded 19.05.2007)

[289] Firmenschrift
GFK(GW)-Rohre im Anlagenbau – Beständigkeit 3 und Mitteilung vom 20.08.2003
Amitech Germany GmbH, Mochau/ Großsteinbach

[290] Strongwell, Bristol (VA/USA)
Corrosion Resistance Guide (Online in the Internet)
<http://www.strongwell.com>
(downloaded 21.02.2008)

[291] Fibrolux GmbH, Hofheim
Chemische Beständigkeit von Profilen und Gitterrosten mit Vinylesterharz-Matrix (Online in the Internet)
<http://www.fibrolux.com>
(downloaded 10.12.2007)

[292] Firmenschrift
Fiberdur® – Lieferprogramm mit Korrosionstabellen, 02/2003
Fiberdur GmbH & Co. KG, Aldenhoven

[293] Firmenschrift
Wavistrong® Epoxy Pipe Systems – Chemical Resistance List, 02/1998
Future Pipe Industries B.V., Hardenberg (Netherlands)

[294] Conley Corp., Tulsa (OK/USA)
Chemical Resistance of Conley Piping Systems (Online in the Internet)
<http://www.pep-plastic.com>
(downloaded 10.12.2007)

[295] Champion Fiberglass, Inc., Spring (TX/USA)
Chemical Resistance of Fiberglass Pipe Resins (Online in the Internet)
<http://www.championfiberglass.com>
(downloaded 10.12.2007)

[296] Oswald, K. J.
Carbon Dioxide Resistance of Fiberglass Oil Field Pipe
Materials Performance 27 (1988) 8, pp. 51–52

[297] Dual Laminate Fabrication Association (DLFA), USA
Chemical Resistance of Thermoplastics used in Dual Laminate Constructions (Online in the Internet)
<http://www.dual-laminate.org>
(downloaded 10.12.2007)

Index of materials

1
1.0037   65
1.0375   53
1.0545   65
1.1003   53
1.1104   53
1.1623   53
1.1632   52
1.4404   89
1.4541   81, 109
1.4571   174
1.5069   63
1.6580   63
1.7220   63
1.7335   42, 65
1.8907   65
1.8961   53
1.8962   53
1.8975   60
13CrMo4-4   65–67
13CrMo4-5   42, 65

3
30CrNiMo8   63, 65
34CrMo4   63, 65
36Mn7   63–65
37Mn5   63
304   90
304 H   90–92
308 H   90, 92
309 L   90
310   109
316   89, 109
316 H   90, 92
316 L   90, 92
321   109
347   109

8
825   95, 185

a
ABS   22, 124, 133–134
AC66   87
ACM   142, 170, 172
acrylate rubber   142, 172
acrylate styrene acrylonitrile
   copolymer   124
acrylic copolymer   148
acrylonitrile butadiene rubber   27, 140,
   157, 171
acrylonitrile-butadiene-styrene
   copolymer   124, 133–134
acrylonitrile styrene copolymer   22
Advantex® glass   175
AFM 34   169
AISI 304   90
AISI 304 H   90–92
AISI 308 H   90, 92
AISI 309 L   90
AISI 310   109
AISI 316   89, 109
AISI 316 H   90, 92
AISI 316 L   90, 92
AISI 321   109
AISI 347   109
Al99,5   31–32
Alkadur® HR   158
Alkadur® K 75   156
Alloy 825   95
AlMg   32
AlMgSi   32
AlMn   32
aluminium   30, 183
aluminium alloys   32, 183
aluminium-boron-silicate glass   175
aluminium materials   30

a

aluminium oxide   106, 108, 186
AM20   92–94
AM60   92–94
Anchor-Lok® PE   166
Anchor-Lok® PP   166
Anchor-Lok® PVC   167
Anchor-Lok® PVDF   167
aramid   170
Aropol® 7241   176
ASA   22, 124
Asplit® CN   155
Asplit® ET   156
Asplit® FN   155
Asplit® HB   156
Asplit® VEC   156
Asplit® VEQ   156
Atlac® 382   176
Atlac® 430   176
Atlac® 580   177
Atlac® 590   176
AU   28, 144, 170, 173
austenitic nickel chromium steels   80–81
austenitic steels   72, 88, 91
AVT 530   177
AZ91   92–94

b

beryllium oxide   106, 186
BIIR   157–159, 162, 189
binding agents   104
Bisphenol resol   152
bitumen   156
Bornumharz® 6101   161
boron carbide   111, 186
BR   27, 163
bromobutyl rubber   157, 159, 162
Bücolit® V25   154
Bücolit® V47-36   153
Bücolit® V590G   152
butadiene caoutchouc   163
butyl rubber   27, 140, 157, 159, 162, 172

c

C15   70
CA   26
CAB   26, 134
cadmium   97
carbides   106
carbon   100–101, 170
carbon graphite   174
cast steel   41, 183
cellulose acetate   26, 131
cellulose acetobutyrate   26, 134
cellulose hydrate   26
cellulose nitrate   26
cellulose triacetate   26, 131
cement mortar   104, 156
cement mortar waterglass cements   158
Centricast® II   178
ceramic materials   109
ceramics   108
cerium sulfide   111, 187
Chemonit® 3B   159
Chemonit® 10   160
Chemonit® 31 HW   160
Chemonit® 34   160
Chemonit® 34HW   160
Chempruf® 120   153
Chempruf® 121   152
Chempruf® 130   152
Chempruf® 131   152
Chempruf® 141   152
Chempruf® 1300   153
Chempruf® 1310   153
Chempruf® 1410   153
Chempruf® 2101   154
Chempruf® 2201   154
Chempruf® 2300   154
Chempruf® 2310   154
Chempruf® 2410   153
chlorobutyl rubber   141, 157, 159, 162
chloroprene rubber   27, 139, 157, 159, 163, 171
chlorosulfonated polyethylene   141, 157, 162, 172
chrome-molybdenum casting   174
chrome steels   67–68, 184
chromium   33
chromium alloys   33
chromium molybdenum steels   67, 184
CIIR   141, 157–159, 162, 189
CN   26
cobalt alloys   32
cobalt-chromium-tungsten alloys known as   32
cold-vulcanizing rubber sheets   162
concrete   101, 103–104, 186
copper   33, 38–39, 183
copper alloys   38
copper base alloys   33
copper-aluminium alloys   40, 183
copper materials   38
copper-nickel alloys   40, 183
copper-nickel materials   40, 183
copper-tin alloys   40, 183
CR   27, 139, 157–159, 163, 171, 189
CrNi steels   89
CSM   141, 144, 157–159, 162, 172, 188

CTA   26
CuNi10Fe1Mn   40
CuNi90/10   40
CW352H   40

d
Derakane® 411   176
Derakane® 470   176
Diabon®   100
Disboxid® water protection system WHG Standard   158

e
EC   26
EC Duro-Bond® Chlorbutyl   162
EC Duro-Bond® E-CTFE Lining   165
EC Duro-Bond® ETFE Lining   165
EC Duro-Bond® FEP Lining   166
EC Duro-Bond® Hypalon®   162
EC Duro-Bond® MFA Lining   166
EC Duro-Bond® Neoprene®   163
EC Duro-Bond® PFA Lining   166
EC Duro-Bond® PP Lining   165
EC Duro-Bond® PTFE-M Lining   165
EC Duro-Bond® PVDF Lining   165
EC Duro-Bond® Rubber   159
ECO   144
ECR glass   175
ECTFE   21, 120, 122, 165, 180–181, 187, 189
Edlon®   150
elastomers   188
electrolytic copper   33
electrolytic zinc   97
enamel   105, 186
EN AW-5052   32
EN AW-6063   32
EN AW-Al Mg0.7Si   32
EN AW-Al Mg2.5   32
Enduraflex VE621BC   162
EP   152, 175, 189
epichlorhydrin rubber   144
EPDM   27, 141, 144–145, 172, 188, 190
EP-Novolak resin   152
Epover®-S   149
epoxy resins   104, 156
EP resins   152, 177, 187
Esshete® 1250   87
ETFE   20, 120, 122, 165, 180–181, 187, 189–190
ethyl cellulose   26
ethylene chlorotrifluoroethylene copolymer   21, 120, 165, 181

ethylene propylene diene rubber   27, 141, 172
ethylene tetrafluoroethylene copolymer   20, 120, 165, 181
ethylene-vinyl acetate copolymer   25, 126
EU   28, 144, 170, 173
EVA   126
EVAC   25

f
factory rubber linings   162
FEP   20, 120, 122, 150, 166, 180–181, 187
ferritic steels   88
FFKM   143–144, 170, 173, 188, 191
Fiberdur® CSEP   178
Fiberdur® CSVE   178
Fiberdur® EP   178
Fiberdur® VE   177
fine-grain structural steels   65
FKM   27, 142, 144–145, 172, 188
floor coatings   158
Fluon®   150
fluorinated rubber   27, 142, 172
fluorosilicone rubber   143, 173
fluorothermoplastics   187
Furadur®-F-Kitt   155
Furadur®-Kitt   155
furan resin   154–155, 190
FVMQ   143, 173

g
Genakor® 022   160
Genakor® 022R   160
GFRP   175, 180, 190
glass flakes   152
gold   32
gold alloys   32
Gore-Tex® Series 300 gasket tape   170
Gore-Tex® Series 600 gasket tape   170
graphite   100, 108, 169–170, 186, 191
Gylon® Blau   169
Gylon® Standard   169
Gylon® Weiß   169

h
Halar®   150
hard rubber linings   161
heat-resistant steels   41
Hetron® 197-3   176
Hetron® 922   176
Hexoloy®   110
high-temperature thermoplastics   122
HNBR   27, 140, 145, 171

i

IIR   27, 140, 157–159, 162, 172, 189
Incoloy® 825   95, 185
IR   27, 158–159, 190
iridium   97
isobutene isoprene rubber   140, 172
Isolast® J9510   174
isoprene rubber   27, 158–159

k

Kerabonit® D3 HW   160
Kerabutyl® BB-S   164
Kerabutyl® BS   163
Kerabutyl® V   163
Keracid® EP 1102   158
Keranol® EP 310   156
Keranol® FU 310   155
Keranol® FU 320   155
Keranol® UP 311   156
Keranol® VE 311   156
Kerapren® VB   164
Klingersil® C-4430   169
Klingersil® C-4500   169
Klingertop-chem®-2000   169
Klingertop-graph®-2000   169

l

laminate coatings   153
lead   97
lead alloys   97
Liquidline® 100   161
low-alloy chromium-molybdenum steels   44
low-alloy steels   41–42, 44, 47, 65, 67, 183

m

MABS   124
Macrolon® 3200   131
magnesium   92, 94
magnesium alloys   92, 94
magnesium materials   93
magnesium oxide   106, 186
martensitic steels   68, 184
Matrimid® 5218   131
MC protection system 1900   158
mechanical seals   174
methacrylonitrile styrene copolymer   23
methyl-methacrylate-acrylonitrile-butadiene-styrene copolymer   124
methyl rubber   27
MFA   120, 122, 166, 180, 187
molybdenum   94
molybdenum alloys   94
molybdenum disilicide   111, 187

mortars   101, 104, 186
MQ   28, 143, 173
mullite   108

n

natural rubber   27, 139, 158–159, 171
NBR   27, 140, 144–145, 157–158, 163, 169, 171, 174, 188, 190
nickel   94, 185
nickel alloy   95
nickel-chromium-iron alloys   95, 185
nickel-chromium-molybdenum alloys   95, 185
nickel-chromium steels   79
Ni-Cr-Fe alloys   95
NiCrMo alloys   96
Nil-Cor® 300   178
Nil-Cor® 310   176
Nil-Cor® 500   178
Nil-Cor® 510   177
Nil-Cor® 610XP   177
niobium   97, 185
nitrides   106
nitrile caoutchouc   140
nitrile rubber   163, 169, 171
novaflon® 200   169
novaflon® 500   169
novaphit® SSTC   169
novaphit® VS   169
novatec® PREMIUM II   167–169
novolak resin   152
NR   27, 139, 158–159, 171, 189–190

o

one-site rubber linings   164
osmium   97
OT   144
oxide ceramics   106
Oxydur® Flake   153
Oxydur® HT   156
Oxydur® TM   156
Oxydur® VE-K   156
Oxydur® VE-L   154

p

PA6   24, 125
PA11   24, 125, 150
PA12   25, 126
PA66   24, 125
PA610   125
PAI   122
Palatal® A 410   176
Palatal® P 69   176
palladium   97

Index of materials

PAN 23, 170
PB 114
PBI 22
PBT 23, 124
PC 24, 125, 127, 133–134
PCTFE 20
PE 133, 166, 180, 187, 190
PEEK 21, 122
PE-HD 19, 112, 115, 180
PEI 21, 122, 134
PE-LD 18, 112
PE-LLD 19, 116
PE-MD 19
perfluoro rubber 143, 173
PESU 21, 121
PET 23, 124, 133–134
PE-UHMW 113, 190
PE-X 113
PF 189
PFA 20, 120, 122, 166, 180, 187
PFA/FEP 189
Pfaudler® PharmaGlass PPG 105
PFTE 189
phenolic resin 138, 151, 154–155, 186, 190
PI 22, 122
PIB 20, 114, 157, 191
pipe steel 62
Plascoat® PPA 571 149
Plasguard® 4550 152
platinum 97
platinum metals 97
PMMA 25, 126–127, 129, 133–134
PMQ 28
poly(2,2,4,4-tetramethylcyclobutane carbonate) 24
poly-2,6-dimethyl-p-phenylene oxide 131
poly-4-methylpentene-1 116
polyacrylonitrile 23, 170
polyamide 6 24, 125
polyamide 11 24, 125, 150
polyamide 12 25, 126
polyamide 66 24, 125
polyamide 610 125
polyamideimide 122
polyaryl ether ether ketone 21
polybenzimidazole 22
polybutadiene 27
polybutadiene, hydrated 27
polybutene 114
polybutylacrylate 104
polybutylene terephthalate 23, 124
polycarbonate 24, 125, 127, 131, 133–134
polychlorotrifluoroethylene 20
polydicyclohexyl siloxane 28
polydimethyl butadiene 27
polydimethylsiloxane 28
polyester resins 137
polyetheretherketone 122
polyetherimide 21, 122, 131, 134
polyethersulfone 21, 121, 131
polyethylen naphthalate 23
polyethylene 18, 112, 115–116, 133, 166, 180
polyethylene terephthalate 23, 124, 133–134
polyethylmethacrylate 25, 130
polyimides 22, 122, 131
polyisobutylene 20, 114, 157
polyisoprene 27
polymer concrete 104, 186
polymer dispersions 148
polymer materials 158
polymethylethyl siloxane 28
polymethylmethacrylate 25, 104, 126–127, 129, 133, 134
polymethyloctyl siloxane 28
polymethylpentene 114
polymethylphenyl siloxane 28
polymethylpropyl siloxane 28
polyoctenamer 27
polyolefine 112–115
poly(oxy-2,6-dimethyl-1,4-phenylene) (polyphenylene ether) 26
polyoxymethylene 26, 126
polyphenylene oxide 126, 133–134
polyphenylene sulfide 121, 175–178
polypropylene 19, 113, 165–166, 180
polystyrene 22, 123, 127–128, 133–134
polysulfone 21, 121, 131, 134
polysulphide rubber 144
polytetrabutylene carbonate 24
polytetrafluoroethylene 20, 119, 165, 186
polytetramethylene carbonate 24
polyurethane 28, 133, 152–153, 156
polyurethane rubber 144, 173
polyvinyl alcohol 25
polyvinylbenzoate 25
polyvinyl chloride 20, 112–115, 119, 133–134, 157, 165, 167, 180
polyvinyl fluoride 21
polyvinylidene chloride 20, 115
polyvinylidene fluoride 21, 121, 133, 165, 167, 181, 186
poly(vinyl-m-methylbenzoate) 25
poly(vinyl-p-isopropylbenzoate) 25
POM 26, 126
Portland cement 101, 104
potassium silicate 156

Index of materials

powder coatings 151
PP 19, 113, 115, 165–166, 180, 187, 190
PPE 26, 126, 133–134
PPS 121, 175–178
pre-vulcanized rubber sheets 162
Proco® – L 151
Proguard® CN 100 152
PS 22, 123, 127, 133
PS-HI 22, 123, 134
PSU 21, 121, 134
PTFE 20, 119, 122, 169–170, 180, 186–187
PTFE-M 165
PU 133
PUR 189
pure aluminium 31–32
pure copper 35, 37
pure iron 47, 57
pure zinc 98
PVAL 25
PVC 133
PVC-C 115, 180, 187, 190
PVC-P 20, 115, 117, 119, 157, 165
PVC-U 20, 114–115, 134, 167, 180, 187, 190
PVDC 20, 115
PVDF 21, 121–123, 133, 165, 167, 180–181, 186–187, 190
PVF 21

r
ramie 170
RBSiC 108
reaction resin putties 155
Rhenoguard® Jumbo 150
rhodium 97
Rilsan® PA 11 150
rubber coatings 157
ruthenium 97

s
S235JR 65–66
S355N 65–66
S500N 65
SAN 124
Sanicro® 28 87
SB 123
SBR 139, 158, 160, 171, 190
SBR, IR 189
sealing layers 156
SiC 106, 108, 175, 186
Sigrabond® 101
Sikafloor® water protection system 390 158
silicon carbide 106–107, 111, 175, 187
silicone rubber 143, 173
silicon nitride 109
silver 30
soft rubber linings 164
spray-applied coatings 151
St 37-2 65
stainless steels 73
StE 355 65
StE 500 65
Stellite® 32
StoCretec WHG System 1 158
styrene-acrylonitrile copolymer 124
styrene-butadiene copolymer 123
styrene-butadiene rubber 139, 158, 160, 171

t
tantalum 97, 185
tempered steels 62, 184
tetrafluoroethylene-hexafluoropropylene copolymer 20, 120, 150, 166, 181
tetrafluoroethylene/perfluoromethyl vinyl ether copolymer 120, 166
tetrafluoroethylene-perfluoropropylvinylether copolymer 20, 120, 166, 180
thermoplastic elastomer 135
thermoplastics 115
thermoplastic webs 157
tin 97
tin alloys 97
titanium 97, 185
titanium alloys 97, 185
titanium carbide 111
titanium nitride 109, 111
TP347H 87
TPE 135
TPE-U 26, 135
Trovidur® W 2000 165
trowel-applied coatings 153

u
Udel® P3500 131
Ultem® 1000 131
Ultrason® E 6010P 131
unalloyed steel C15 70
unalloyed steels 43, 51–52, 56
unsaturated polyester resin 156
UP 152, 175, 187, 189
UP HET acid resin 152
UP resins 104, 176, 186

v
VE 152, 175, 187, 189

VE resins   176
vinyl ester resins   138, 156
VMQ   143, 173
Vulcoferran® 2190   160
Vulcoferran® 2194   160
Vulcoferran® 2208   162, 164
Vulcoferran® 2503   163
Vulkodurit® 1250   160
Vulkodurit® 1691   162—75
Vulkodurit® 1755   162
Vulkodurit® D3   159

W

Wagulast® BIIR 1642   162
Wagulast® BIIR 1643   164
Wagulast® CR 1504   164
Wagulast® CSM 1717   164
Wagulast® CSM 1720   162
Wagulast® NBR 1833   164
Wagulast® NBR 1842   163
Wagulast® NR 1358   163
Wagunit® 1050   160
Wagunit® H 1000   159
Wagunit® H 1010   160
Wagunit® H 1110   160
Wagunit® H 1118   159
Wagunit® H 1122   160
Wavistrong®   178
welding materials   90

X

X10   87
X20   87
X20Cr13   69–71
X 65   60, 62

Z

zinc   97
zirconium   99, 185
zirconium alloys   99, 185
zirconium dioxide   106, 186
zirconium oxide   108

Subject index

a
aging resistance   159
anhydride of carbonic acid   5
aqueous carbon dioxide-containing solutions   96
aramid fiber   169
ash deposits   87–88, 185

b
beverage industry   33
binding agent   186
biomass   87
Boudouard reaction   79
Boudouard's equilibrium   43
bound carbonic acid   11
bubble formation   145

c
carbide formation   16
carbon activity   15–16
carbonate hardness   11
carbonatization   101–103, 186
carbonatization coefficient   102
carbonatization depth   104
carbonatization rate   104
carbon diffusion   95
carbon fibers   169
carbonic acid excess   11
carbonic acid snow   5
carbon monoxide   95
carburization   **16**, 95, 185
carburizing gases   95
catastrophic carburization   16
CERT tests   66
C-glass   175
chemical and petrochemical plants   72
chemical equipment   180
chemical plant construction   33
chemical vapor deposition   106
cladding   29
coatings   29, 148
cold forming   56
cold water pipes   38
combined linings   155
combustion gases   72, 87
compression heat   14
compression packings   170
condenser   33
copper components   40
copper corrosion   37
copper oxide layers   40
copper pipes   33, 35
corrosion   17
corrosion protection   29
cracking tendency   64

d
device flange connections   170
dissociation constant   6
drinking water distribution systems   38
drinking water pipes   33
drinking water supply   39
dry ice   5, 14
duroplastics   187

e
E-glass   175
elastomer webs   158
electrochemical measurements   52
enamel coatings   105
erosion corrosion resistance   70
ethanolamine process   14
excessive free carbonic acid   11
explosive cladding   29
explosive decompressions   145, 174, 188

Subject index

f
factory rubber linings   158
fillers   104
fire extinguishers   63
firmly bound carbonic acid   11
flake filler   152
flat seal   167
flow rate   44
flue gas purification   14
free carbonic acid   11
fresh water pipeline   33
fuel element cladding tubes   79

g
gas bottles   63
gas containers   65
gas pipelines   60
German Washing and Cleansing Agents Act   13
glass fiber reinforced plastics   175
glass transition temperatures   117, 127, 130
grain boundary sensitization   90
graphite   16
graphite plates   106
gratings   177
green rot   95

h
hard chromium layers   33
hardness value   13
hard rubber linings   158, 189
hard surfacing layer   90, 92
heat exchangers   33
heat exchanger tubes   41
high-pressure vessels   145
high-temperature corrosion   107
hydration   51
hydrocarbons   95
hydrogen embrittlement   16
hydrogen evolution   51
hydrogen-induced crack formation   65
hydrogen-induced stress corrosion cracking   16, 62, 65
hydrogen permeation measurements   65
hydrogen sulfide   15
hydrolytic attack   17

i
industrial atmosphere   92
industrial floor coatings   158
inhibitors   38

l
laminate coating   189
laminates   176
lime-aggressive carbonic acid   11
lime-aggressive (lime-attacking) carbonic acid   11
lime-carbonic acid balance   **11**
limestone   12
linings   148
loose jacket lining   189

m
magnetite   41
magnetite layer   41–42
manufacture of chemical equipment   115, 187
maritime environment   92
medical technology   33
metal dusting   **16**
mineral fillers   161
mineral flakes   152
mineral waters   40
mullite layers   109

n
natural gas production   62
nuclear reactors   41, 79

o
oil extracting facilities   15
O-rings   167, 170–174
oxidation   **16**
oxygen activity   15

p
packings   167, 170
peposition welding   29
periods of downtime   40
permeability   104, 116, 118, 123, 149, 159
permebility coefficient   18, 115–119, 123, 127–131, 135, 145–146, 148
permeation coefficients   117, 135
permeation   18
petrochemical plants   57
petroleum production   15
pipes   177
pitting corrosion – type 1   38–39
plasma spraying   109
plastification pressure   131
pore volume   104
powder coating   189
power plants   72, 87
processing of crude oil and natural gas   47
process waters   97

Subject index

production   69
protective film   29

r
related free carbonic acid   11
reservoirs   47
retention ponds   158
Reynolds number   70
roll cladding   29
rotating disk specimens   53
rubber linings   148, 158
rubber powder   161
rubber webs   158
rural environment   92

s
salt melts   108
salt spray tests   93
scale   12
scale-offs   41
scaling   42
sealants   145
sealing rings   170–173
sealing strip   170
seawater desalination plant   40
silver solders   30
sliding ring seal   167
slow strain rate tensile tests   66
soft rubber linings   159, 189
solid particle   71
solubility coefficient   8
sorption   146
sour gas condensate   65
sour gas plants   32
spray-applied coating   189
spraying process   161
sterilization methods   132
stress corrosion cracking   63, 184
stresses   17
sublimation temperature   5
swelling   17
syngas plants   65
synthetic drinking water   37

t
tanks   158
technical solubility coefficient   7
temperature resisting steel   42
test procedures   28
textile glass   175
thermal conductivity   5
thermal shock resistance   96
thermal spraying   29
thermoplastic elastomers   187
thermoplastic linings   189
thermoplastics   187
thin coating   189
titanium nitride layers   109
total carbonic acid   11
total water hardness   13
transcrystalline stress corrosion
   cracking   65
transport of natural gas and crude oil   68
triple point   14
trowel-applied coating   189
types of damage   17

u
ultrapure carbon dioxide   54

w
water aggressiveness   12
water/cement ratios   102, 104, 186
water hardness   12
water pipe systems   **38**
water pressure test   63
water protection   158
water's degree of hardness   12
water vapor content   42
water vapor permeability   148–149, 164, 187
welding technology   33

y
yarns   170